# THE INTRICACY GENERATOR

# THE INTRICACY GENERATOR

## Pushing Chemistry and Geometry Uphill

Peter G Kinnon

**PICKLE PUBLICATIONS**

Pickle Publications is a trademark of

Peter Kinnon Literary Enterprises



National Library of New Zealand Cataloguing-in-Publication Data

Kinnon, Peter G. (Peter George), 1938-
The intricacy generator : pushing chemistry and geometry uphill / Peter G. Kinnon.
ISBN 978-0-9582876-2-3
1. Evolution (Biology) I. Title.
576—dc 23

## CONTENTS

**APPENDICES**

# Preface

This book will displease many.

It will displease those who are resolutely committed to any kind of religious world view because, although the historical importance of religion and other superstition in the evolution of technology is brought to attention, there is no attempt to reconcile religion with any aspect of current science. Or, indeed, in the light of our modern understandings, to accept that such myths have any further useful part to play.

It will similarly be found distasteful to some of those within the scientific community who are saddled with institutional dogma. Specifically, those whose minds reflexively snap shut like a clam at the merest whiff of anything remotely suggestive of kind of directionality. Even though the very effete form of teleology implied here owes nothing to the concept of a deity. and a great deal to the concept of a ratchet!

It may even upset some of the more open- minded and more rational scientists, because science is treated herein, not without good reason, as a sideshow of technology.

While that scenario might well be welcomed by the engineer, who sometimes could justifiably grouse at the lack of public recognition for his field and against science getting all the "glory". He, in turn, might rail against the proposition that, except in a trivial everyday sense, there are really no designers. The philosopher, unless one of the few that work under the system of axioms which corresponds to the empiricist faith, will find the book far too prosaic, pedestrian and free of ontology and obfuscation to stir his spirit.

It will inevitably annoy some biologists of the old school who are still ideologically bound to the gene-centred paradigm of inheritance. For their beloved DNA, sometimes wrongly treated as the "brains of the cell" is here implied to instead play a rather small part.

The layman, in common with many of the above, also may take umbrage at the proposition that our species is characterised not by the vague woolly notion of "intelligence" but rather by the complex neural mechanisms that serve the hands and the vocal apparatus. Even though this inference is very well founded.
Inevitably, the somewhat unorthodox interpretations promoted herein will be brushed aside by many such. This, of course, is only to be expected. In the words of Arthur Schopenhauer:

> *"All truth passes through three stages. First, it is ridiculed. Second, it is violently opposed. Third, it is accepted as being self-evident."*

On the positive side:

In contrast to my two previous works, this more formal book is aimed at the reader who has either a working background in the basic sciences or has the motivation to acquire such knowledge. Which, in today's information-rich environment, is not at all difficult.

At the same time, it is my intention to keep the treatment sufficiently straightforward for use in colleges. As ancillary reading that provides a broad framework which, in conjunction with "The Goldilocks Effect", can serve as a launching platform for encouraging interest in all manner of scientific pursuits. And also for developing the habits of creative and critical thought. Perhaps by picking holes in my reasoning, which of course, may be flawed. After all, although I hope to have discharged my responsibility of due care, I can make no guarantee of its integrity. To echo the sentiment once expressed by Michael Faraday, I only hope that my thoughts will be found worthy of consideration.

Except for the brief extrapolation into the future towards the end of the volume, the tangential speculations in appendix A regarding a non-arboreal prehistory of our species, and the involvement of warfare in the explosive cerebration event that occurred in human evolution, the discussion is entirely based upon well-established mainstream scientific principles and observations. Along with some valuable contributions from the broader study of history. Which is most certainly not to be dismissed as "just one damned thing after another". Rather, the book is, for the most part, simply an attempt to encourage an interpretation that does not rely upon the anthropocentric world-view with which we are surely burdened by dint of our genetic and cultural legacies. A set of prejudices and conceits which can seriously cloud our objectivity.

There are encouraging signs that the emerging new generation of chemists, biochemists and particularly astrobiologists are breaking away from the anthropic prejudices to which many of their elders, one way or another, are bound. This book aspires to both support those riding this exciting new wave and also to provide inspiration to youngsters soon to embark on their careers.

I make no apology for iconoclasm. Although there are several examples of it in this book they are not motivated by any kind of malice. But idols, idol-worship, and sheep-like uncritical acceptance of authority deserve no place in the sciences.

On a practical point, it will be noted that I have broken with the tradition, increasingly common with popular books of this kind, of providing references. There are several reasons for this:

- There is a growing tendency among writers to provide such lists out of all proportion to the original content of the work.

- Where such exhaustive lists are included, the main object sometimes appears to be to impress and to "pad out" the volume rather than to substantially increase understanding.

- Quite understandably, even when used for the original purpose, there is often a considerable degree of "cherry picking" of references that serve the contentions of the author.

- They have the potential to disrupt the flow of arguments being advanced.

But, most importantly, for this kind of work at least, they have today become redundant. Because we now have instant search capabilities at our fingertips that provide the great advantage of raising contrary arguments or evidence in addition to those that give support.

The style that I have adopted for this work is unorthodox. The brief italicized vignettes interposed between chapters are meant to encourage readers to navigate the content according to their own particular interests. Intended primarily as "jump-points", they may also provide a convenient way of initial "skimming" to get the overall drift before reading the chapters in their entirety.

Another departure from the norm is the fugal structure whereby the basic themes are woven contrapuntally through many chapters. While I have arranged these in basically chronological order, I have also endeavoured to make them to some extent self-sufficient. This inevitably involves some degree of repetition, but where this occurs each iteration is coloured by a rather different perspective. A feature which, I hope, might help to minimize misunderstandings.

# Introduction

The predominant theme of this discussion is to draw attention to the evolutionary continuum which has its roots in the formation of stars, extends right through to the present era, and will presumably continue indefinitely into the unforeseeable future. Most importantly, it recognizes that the evolving entity is essentially that of atomic species and their subsequent combination into a population of novel molecules and assemblages thereof, of ever increasing overall structural intricacy.

This evolutionary pattern is very distinct. Starting with the pre-stellar atomic species of atomic number one, plus a small proportion of helium and a tiny amount of lithium, we have now reached, in our own little corner of the universe, a situation where the number of compounds listed by the Chemical Abstracts Service is approaching sixty million, with twelve thousand new substances added every day. Also, it is accompanied by a complementary increase in intricacy of the geometry of their assemblages. Some examples are the solar systems of astronomy, the cell of biology, the airliner and the suspension bridge of engineering. The lithographically generated integrated circuit of electronics.

This persistent and all-pervasive pattern of a self-propagating system that has exhibited strong directionality over the aeons is not to be taken lightly. Neither is there the slightest justification for interpretation in terms of superstitious mythology for which there is absolutely no evidential basis. Nor is it to be carelessly swept under the carpet by equally unevidenced speculations of a multitudinous population of universes, each having different physical properties.

Quite apart from the lack of evidence, or even direct testability, such conjectures are contrary to the principle of parsimony which, while admittedly not infallible, has served us very well as a good rule of thumb in the progress of scientific endeavour.

The same lack of parsimony applies to those covert religious apologists who accept biological evolution along with the main body of science but try to use the observed directionality in an attempt to justify the notion of "intelligent design". It is of course, quite valid to maintain compatibility with science by "plugging in" a deity at a sufficiently early point in the history of the universe. Such as "before" the hypothesized big bang. But to do so is entirely unnecessary. For, as we shall see later (at the expense of some human conceits), the observed directionality provides no such justification.

Various authors, typically proclaiming the discovery of "order arising from chaos" have been hornswaggled into the platonic belief that this widely observed pattern corresponds in some way to the superficially intricate patterns that are able to be generated by iteration of mathematical expressions of the kind that, when appropriately processed for display, give rise to fractal patterns. Such as, by way of example, the "Julia set". In actuality there is no sound basis for this assumption.

Firstly. it is important to remember that these have no independent physical existence outside the computer system in which they are generated. Certainly they can be (partly) exported by, say, printing. But the print-outs are sterile. They have no potential for self-replication. The morphologies of products of the life process do ***not*** exhibit the scale invariance of fractals. Trees, lungs and rivers are among the examples often claimed. While it is true that, depending upon the overall size of the network, gross similarities may be observable over many orders of magnitude, these branching networks merely reflect, in the case of systems such as lungs and trees, optimal surface area access. While the branched structure of rivers is, of course, entirely determined by gravitational potential, the volumes of water involved and the terrains over which the water traverses.

In these, and in all other claimed instances of fractality in nature, the self-similarities are only very approximate and the scaling is limited.

Secondly, although a feedback loop is involved in their generation within the computer system they have no capability for interaction with the external environment.

Thirdly, and most importantly, while superficial intricacy can be said to be generated by algorithms of this kind, no functionality or even continued structural diversity, other that of scale, results from the output.

The fallacies of this kind associated with "chaos" arise largely as a result of a conflation of the traditional concept, which is related to randomness, with the modern technical use of the term in connection with the field of study that involves the exquisite dependence of the progression of non-linear systems upon initial conditions..

Others have pointed to iterative logical algorithms such as this Commodore Basic program line as a generator of intricacy:

```
10 PRINT CHR$(205.5+RND(1)); : GOTO 10
```

This, when output, produces a complex "maze". In this instance we see again that no "order arises from chaos". Indeed, outside the constraints of the system, no order is really produced at all. The configuration of the resulting display is random with no consistent functionality or structural diversity

There is no justification for embarking upon such unnecessary quasi-metaphysical speculations as these when, by keeping our feet firmly on the ground and simply using well-evidenced scientific knowledge in conjunction with basic engineering principles and, where further required, a smattering of heuristics, we can "connect the dots" to draw a quite comprehensive, if necessarily incomplete, picture of the ever- increasingly intricate universal network .

An evolving entity which, by stealing a term usually applied to biology, we can usefully equate with “life”.

The "order out of chaos" notion seems to stem from attempts to rationalize a remnant of superstitious mythology. The universe we observe, although having a very considerable probabilistic component, and being very far from equilibrium, is certainly not chaotic. Nor is there the slightest direct evidence that it ever was.

The subtitle of this book "Pushing Chemistry Uphill" alludes to the role of gravity, as the prime motivator of this process, persistently overcoming what is generally perceived to be a strong statistical tendency for our universe to become globally less structured. At first glance, there are components of this process which might seem not to be aspects of chemistry, or of its primary motivator, gravity. However, on closer inspection we find that this position might not be very far from the truth.

In our examination of the development of intricacy we are largely concerned with the more parochial issues relating to planet Earth. But we should not overlook those large-scale features which, too, exhibit a considerable degree of structure. A rough outline of these, and more importantly, the inter-relationships between the members of this cosmic family gives us chapter three. While even the most teleophobic of individuals would be very hard pressed to deny the inevitability of the gross outcome of nucleosynthesis in, say, a star of solar mass 10, they may still balk at the slightest hint of an overall directionality, even as merely represented by the functioning of components of evolutionary machinery. Chapter 19 will attempt to allay such fears

.

A separate, but not unrelated issue to be explored is the explosive expansion and rearrangement of the human central nervous system during the transition from pre-human to human. While there is a paucity of hard evidence to account for this, we can fairly confidently infer, using our modern understanding of basic engineering principles, that a high level of positive feed-back would necessarily be involved. This being a very well established characteristic of exponential growth.

As well, we would choose as likely contenders for drivers of this process, behavioural aspects which are unique to our species. The use of tools is perhaps the most obvious candidate, but primary tool-use is quite widespread among vertebrates, particularly the primates and the birds. However, secondary tool-use, the ability to use tools to make similar tools, and tertiary tool-use, using tools to make other kinds of tools certainly seem to be among the defining features of the human animal and its Paleozoic predecessors.

The commonly held attribution of social interaction comes some way to meeting this requirement but is not sufficient in itself After all, there are many examples of social animals, none of which have undergone neural transformation that has enabled the generation of a vast range of artefacts or large-scale sharing of imagination.
There is another very pronounced characteristic of our species which is almost always overlooked in this context, but is certainly worthy of consideration. That is the various kinds of "arms race" promoted by our marked predilection for intra-species warfare.

A fourth contender, intimately related to these primary motivators, especially that of tertiary tool use, is the co-evolution of language and greatly extended imagination. Once that process was under way, the level of feedback provided by such interaction leading to the particularly intricate neural hardware that serves the network of minds and allows of the great pool of shared imagination that we have today.

With the establishment of this shared human imagination the scene was set for a nascent phase of the evolutionary process to prosper. The evolution of ideas within this medium of shared minds. Specifically, the autonomous evolution of technology, which now gives every indication of culminating in the emergence of yet another phase of the overall evolution of intricacy and geometry arising from what at present we call the internet.

This book looks at the broader evolutionary process, the increase in chemical and geometrical intricacy, as a whole. Or at least that part which corresponds to observed phenomena. Because of this, many different aspects of this evolving network are put under the spotlight. Chapters are therefore addressed to quite diverse issues, which include formation of the elements, the evolution of minerals, of biology, of our species, of language, and of technology.

Other subjects which are closely related and pertinent to the discussion are the nature of consciousness, the importance of shape, the illusion of "design", and a clarification of the concept of information.

Above all, that the interactions between all facets of the machinery of nature are widespread and part of an integrated evolving network in which the whole can be considered to be more than the sum of its parts. For this reason a systems engineering perspective, which relates to the "big picture", is ideally required. "The Intricacy Generator" is intended to encourage at least a small step in that direction.

# 1

# THE FIRST FAMILY OF STARS

## Emergence of elements 2 to 26

The White Rabbit put on his spectacles. "Where shall I begin, please your Majesty?" he asked.
"Begin at the beginning" the King said gravely.

As the beginning is, in this case, unknown and probably unknowable, we can rather conveniently choose that time in the development of our universe which corresponds to the dawn of the stelliferous era. A time when our universe was entirely populated by just a few neutral atomic species that are believed to have evolved from a "Big Bang" event. Predominantly molecular hydrogen with some lithium and a trace of beryllium.

For reasons not entirely clear, the distribution was not homogeneous but in relatively dense clouds which lent themselves to stochastic clumping, most likely by short-range intermolecular attraction (van der Waals forces). These clouds eventually coalescing into stars of greatly varying mass, some becoming sufficiently dense in their core to cause the initiation of nuclear fusion reactions.

To summarise the stages in this phase of the generation of intricacy:

- Gravity causes this primordial matter to further coalesce into proto-stars.

- Within the resulting accumulations gravitational potential is converted to heat/pressure.

- Heat/pressure and the concomitant intimate proximity of hydrogen atoms provide suitable conditions for initiating the generation of species with higher atomic number.

- In the case of stars around 8 to 20 times the mass of our sun, elements up to the size of iron are produced by gravitation powered fusion

The evolution of these new species is constrained by available pathways for each step. These pathways are provided by the current prevailing conditions, these in turn also being a result of the evolutionary process. It exhibits a very high degree of directionality. We can safely assert that it is certainly not a random process because it follows a clearly identifiable sequence of events.

That is not to say chance is not a significant factor in the progression, but rather that the ratcheting forward of the assembly of species of higher atomic number is at each point dependent upon the selection provided by the currently available states.

New stars are still being formed and because we are able to observe stars at all stages of development and determine their composition spectroscopically, the inevitability of gross outcomes has a well established empirical basis.

# Endothermic formation of elements 27 to 92

For those of the cosmic family having mass comparable that of our sun there is a long pause of many billions of years before formation of the heavy elements can proceed and, when it eventually does, at the stage of the red giant, it is slow and incomplete.

While a less effective provider of atomic intricacy, such stars may, however, while on the main (helium forming) sequence, have the capability to support the later evolutionary phenomena of biology and technology. At least, that has certainly been the case with our own Sun.

The larger stars, in contrast, achieve the formation of iron at their cores within only millions of years. The exothermic process of fusion in the core now has nowhere to go. The gravitational collapse of the core can no longer be kept at bay and occurs suddenly.

This final collapse causes the core to become unstable such that most of the energy therein is released explosively. Providing nuclear fragments that use the excess energy released to allow endothermic assembly of the remaining elements, right up to uranium, to occur.

Again, we have a process that, at small scales, is predominantly random, but constraints of the prevailing environment at any point in time provide directionality and an outcome that is, in some gross sense, effectively predetermined. It is an ubiquitous pattern which we will find repeated in many other aspects of the machinery of nature.

# 2

# DESIGN

*If there was a prize for the word most responsible for the woolly thinking found in discussions such as this it is "intelligent". Along with its equally ill-defined, introspective and often emotive derivatives it is guaranteed to muddy the clearest of intellectual waters.*

*An attempt at its assassination and replacement with the far better term, "imagination" will occur in chapter 27*

*The seemingly far less innocuous word "design", too, has a serious problem. Particularly when used as a verb or in the context of a "designer".*

*It is beyond question that this particular concept is useful in our everyday lives. But if we look more carefully, and from a wider perspective, we find no real evidence for underlying phenomena that correspond to design or designers.*

*The many systems and artefacts that we humans generate are embodiments of ideas. The artefact to the idea is roughly as the phenotype is to the gene.*

"Roughly" because there is no definable unit idea. Rather, as we will find in chapter 27, the neural metaphors that represent the external world are idea complexes of ill defined magnitude.
Furthermore, for the reasons given in chapter 11, the phenotype is by no means fully specified by the gene.

Moreover, no inventions are "drawn out of thin air" as some would have us believe. Rather, they arise from cumulative associations and recombinations of idea complexes. Augmented, of course, by new data gleaned by our senses from the environment. Observations!

So, when we get down to the nitty-gritty, all that remains of the "designer" is a place-holder for evolutionary processes.

The juxtaposition of these two words, one so vague as to be almost meaningless and the other to describe a non-existent phenomenon, is most inopportune.

And the proponents of "Intelligent Design" who have only one apparent real world straw at which to clutch, find it crumbling to nothing at their touch.

Or, perhaps, even more to their distaste, find it to be yet another facet of the very evolutionary process which they despise. Chapter 17 will further demolish "designers".

# 3

# THE COSMIC FAMILY

The variation of mass of the coalescing clouds in itself produces a whole family of celestial bodies. These include the still-born brown dwarfs, having insufficient mass to support hydrogen fusion, stars of masses comparable to that of our sun, for which helium is the limit of the fusion process, and the more massive stars for which iron formation is the exothermic limit. The ensuing type II supernova completes the evolution of the full complement of "naturally occurring" elements that we find today. Accumulations of even greater mass ultimately wind up as neutron stars, and still greater systems . According to the prevailing wisdom, as singularities, "black holes". To what extent these and the many other types play any further roles in the overall evolutionary process we are tracing here and which are mere "spin-off" is hard to judge. As is to what extent the whole system should be viewed as a network. Certainly we should give serious consideration to looking at a galaxy that way. Very recently it has been discovered that galaxies in the vast and mostly uninhabited regions of the universe are aligned into a vast "Cosmic web" of gaseous channels. criss-crossing and interconnecting galaxies. So there seems to be an indication of structured connectivity even at that great scale. Indeed, this network of intergalactic matter is believed to comprise most of the normal matter in the known universe, and to be the cradle in which new galaxies are born.

It is in second generation stars, such as ours, which have coalesced from the element-rich supernoval clouds that we can trace, in at least this one instance, a continuation of the increase in intricacy of the geometry of atomic species and assemblages thereof. While the formation of planetary systems is one of those areas of which our understanding is very sketchy, the sheer number of these suggests that our particular planet may not be alone in supporting the evolution of biology and even of technology. A recent California Institute of Technology study of the planets of Kepler 12 (which they claim to be representative of most stars in our galaxy) suggests that at least 100 billion planets populate our galaxy. And at the University of California, Berkeley, a careful study by a team headed by Geoffrey Marcy, using the Kepler data, similarly led to the conclusion that small planets, with sizes quite similar to our Earth, are, in fact, very common. In Marcy's own words:

> *"You can imagine travelling outward from our solar system, stopping at the nearest sun-like star, then stopping at the second nearest, and so on. When you arrive at the fifth, you have passed five sun-like stars. On average, one of those five sun-like stars has an Earth-size planet in its habitable zone, with lukewarm temperatures, suitable for life."*

These sentiments are echoed rather forcibly by Seth Shostak, who is a senior scientist at SETI.

> *"In the last 50 years, evidence has steadily mounted that the components and conditions we believe necessary for life are common and perhaps ubiquitous in our galaxy. The possibility that life has arisen elsewhere, and perhaps evolved intelligence, is plausible and warrants scientific inquiry. If you extrapolate on the planets they discovered, there are a trillion planets in the galaxy. That's a lot of places for life. We know the majority of stars have planets, but what fraction of stars have planets that are more like the earth? It might be one in five."*

Furthermore, recent work by Linda Elkins-Tanton at Oxford University draws attention to the ubiquity of water in our universe and to the virtual inevitability of oceans re-forming quite quickly after the occurrence of major heating events. She derives an approximation of the initial water content of the earth that arose by the accretion of planetesimals by means of the compositions as represented by zircon found in meteorites. The age of these can be accurately determined by the extent to which the uranium inclusions have been fissioned to lead. The hydration that has been thus observed in these ancient zircon formations indicating a sufficient source of water to provide more than enough to form our oceans by out-gassing from the interior and subsequent re-condensation on encountering the very low temperatures at altitude. Therefore, it would increasingly appear that the prospects for the existence of planets in the "habitable zone" of like-size suns, and the likely presence of oceans on these, are rather good. Also that a serious loss of water by heating, either during formation or as a result of an event as catastrophic as the quite widely accepted moon-forming collision, is very unlikely.

Wherever we have water, we necessarily have preserved a source of two of the other prerequisites for biology, oxygen and hydrogen. The other crucially important elements carbon, sulphur and phosphorus are not volatile. Even the most bioavailable compound of carbon, the CO2 molecule is sufficiently heavy not to easily find, even at escape velocity, a pathway through the busy atmosphere of heavier species at, say, an Earth temperature of 1200C We can reasonably guess that because, for a star to have rocky planets at all its history, it must include a condensation from a supernoval event so as to provide sufficient "metallicity". The accretion of a similar size planet in what is to become the habitable zone, is quite likely to contain the other elements in somewhat similar proportions to that of Earth. For sure, the prerequisites for increases in the intricacy of atomic assemblages of this order are not trivial. But we must bear in mind that our observation of nature's evolutionary machinery shows a profligate disregard for economy. As, closer to home, is evidenced by the billions of sperm sometimes released for the production of a single gamete.

Again, an atom on the sun writes its jiggly energetic signature upon a winged messenger that, ever wiggling, translocates through space-time until that package encounters a recipient atom or molecule by whom its invitation to dance can be read. Or, if of lower frequency, can induce an oscillating electron flow along a conductor. Most of such messages, of course, are never read, but just wind up in the entropic spam-folder of space. The few that find a recipient might give you a nice tan, or melanoma. Or, more commonly, by dancing to a different rhythm, contribute to the evolution of chemical intricacy via photosynthesis.

# 4

# PURPOSE

Time and time again our cogitations and discussions upon the workings of nature fall foul of the vagaries of verbal vagueness. Reason seriously subsumed by semantics.

A lack of precision is inherent in natural language, a reflection upon its flexibility and great richness, and a consequence of its evolutionary history. Our interchange of imagination being highly dependent on contextual cues, which, particularly when engaged with other than mundane matters, can lead us seriously astray.

More straightforward issues allow of being stripped to the bare bones, which some then seize and run for the shelter of the hills where the far less complex language of mathematics is the lingua franca.

But for handling the great complexities of biology, for instance, we are stuck with all the ambiguities of natural languages and need to carefully clean up words, to improve their specificity so we can more easily pick our way through the tangled undergrowth of meaning.

So what can we say about the word "purpose". It is very familiar, of course. Too damn familiar, in fact! We take it for *granted and seldom stop to examine its meaning. We all have an intuitive sense of purpose, which we derive (dangerously) by introspection and also in terms of our own behaviour.*

*The quite distinct empirically derived use of the word suffers from no such difficulties. I suspect that even the most zealous denier would agree that the primary purpose of a spanner is to loosen a nut. Or that anybody would seriously envisage a tool of that kind being endowed with an* ***intent*** *to perform its function.*

*Although some would no doubt try to refer it to the medium of human imagination in which it evolved. In a desperate attempt to bring it back within the traditional introspective domain within which they are, by force of habit, more comfortable.*

*Harking back to the consideration of biology, we objectively observe abundant evidence of purpose in the behaviour of other folk. And that very reasonable interpretation of goal-seeking is reinforced by imagination transfers received from others. By way of speech or of text, for instance.*

*We observe this behaviour in other animals, too. Few would deny purpose to the tiger stalking a deer, or a cow heading for the water trough. We can find purpose, goal-seeking behaviour, in a plant seeking optimal exposure to the sun.*

*We find it on the microbial world as well, in the flagellate bacteria,. These organisms have a bi-directional rotary motor which drives a set of whip-like filaments. When in drive mode the filaments form a coherent bundle and act as a propeller. When the motor reverses, the filaments become disentangled and the bacterium tumbles randomly. This strange intermittent mechanism, however, is nevertheless clearly endowed with purpose. Implemented by sensing a nutrient concentration in each drive mode session and comparing it with the previous. Setting a "switch" accordingly so as to bias its otherwise random motion to find its way to the nutrient source. A very nice instance of a stochastic ratchet in operation.*

*To tie all of this purpose which is inherent in biology into the "big picture" that is provided by interdisciplinary knowledge, we can well ask how it might have arisen? A question to which biological evolution by natural selection provides a very convincing answer.*

*We can also ask ourselves what is the defining characteristic of all this purpose? Very clearly it is directionality. Our motile bacterium is ratcheting up a chemical gradient. The cow is heading for its trough. Both exhibiting directionality. Neither having an "ultimate goal". There will be other gradients, other times to visit the trough. The only goal that they share is that of self-replication, and that is more properly attributed to the overall evolutionary process itself. The turning of the cogs of nature's machinery. Whose only "goal", as far as we can tell, is that of ever greater intricacy.*

*Bear in mind that our bacterium is not endowed with a homunculus, a little humanoid "guy in its head". Neither is the cow. Nor, despite the claims of the religious, is nature. Nor, for all the voluminous candy-floss generated by over-indulgence in self-referential introspection, are we.*

*Some of the features that enable this ratcheting process are discussed in chapter 19.*

*Chicken does not revert to egg*

*Scrambled egg does not un-scramble*

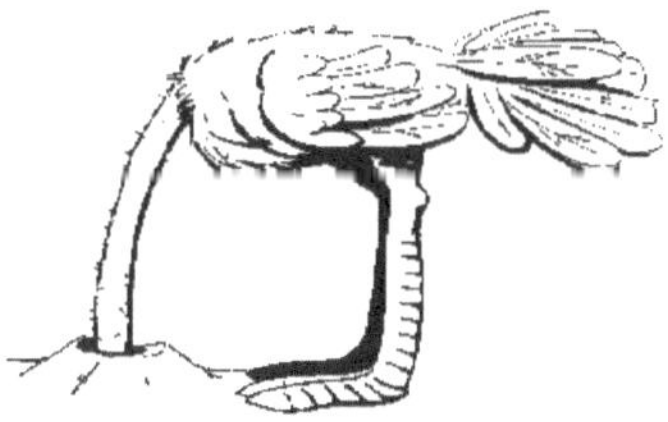

***Distinguished science professor in denial of directionality***

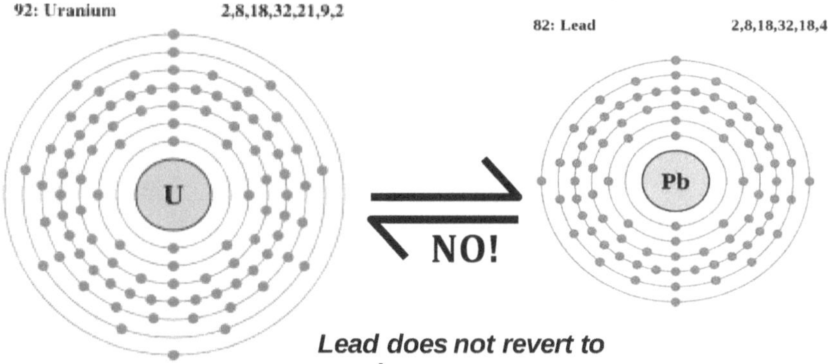


*Lead does not revert to uranium*

# 5

# SHAPE, ORDER & ORGANIZATION

The alert reader may have noticed that subtitle of this book "Pushing Chemistry and Geometry Uphill" can be held to have a redundant term. Because chemistry can be argued to be defined by geometry. Even at the level of the atom we find a distinct shape for each species. Even in quantum mechanical descriptions the "clouds" which represent the probability of finding an electron have distinct, if tenuously bounded, shapes and sizes.

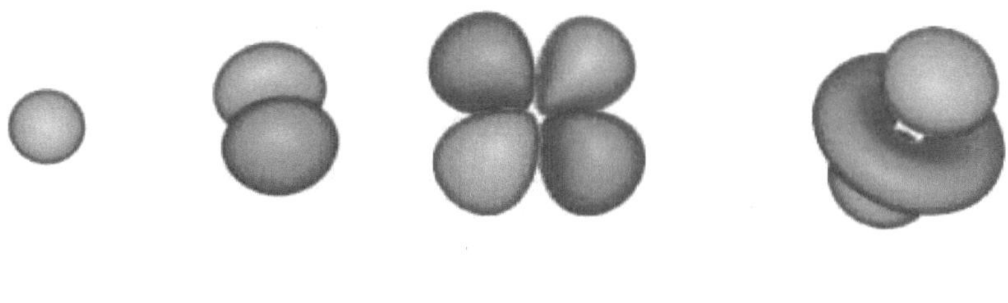

The shapes of molecules inevitably constrain their interactions and largely determine some of their bulk physical properties. Such as viscosity, modulus of elasticity and density. Similarly, the properties of a crystalline solid are dependent on the geometry of the crystal. Graphite, with a planar hexagonal geometry that permits of delocalised electrons, bears no relation in its properties to the cubically arranged diamond.

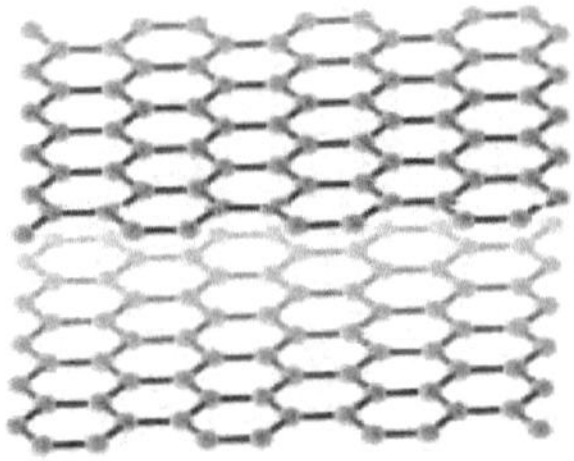 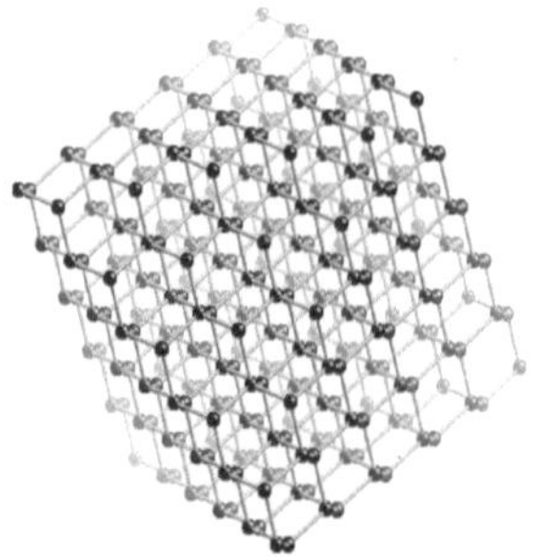

The shapes into which proteins are folded is of profound importance in biological systems. Catalysis in general is largely shape-dependent.

At the other end of the scale we find galaxies. Some, such as that in which we live, have spiral geometries, others are elliptical, some ring-shaped. There are also some that have no discernible regularity but, nevertheless, on the whole we find that, even at the galactic scale, many highly ordered systems have emerged from the relatively homogeneous simple gases of the pre-stelliferous era. And, as mentioned previously, astronomers have found that, even at the largest observable scales, gaseous filaments form an interconnecting web between galaxies. Wherever we look, we find shape, geometry, chemistry and function arising from the interaction of gravitation with this primordial substrate. Ever increasing the intricacy of our universe. No more so is this than in the case of biology, which, indeed, is characterized by its immense proliferation and diversity of shapes. Shapes at once defining and being defined by function.

Order is a concept which in everyday life seems pretty straight-forward. But when discussed within the context of physics and chemistry much confusion and contention can arise. The reason for this is that two quite distinct concepts are usually conflated. Intrinsic order and extrinsic order. Extrinsic order is that which we might perceive in a pack of cards. Typically, if we open a new pack we find the ace of spades first, followed by the two of spades and so on ascending in rank to the king of spades, the cycle repeating through clubs, hearts and diamonds. We intuitively consider this pack to be ordered. But suppose we re-arrange the pack in such a way that the aces of the respective suits come first, then the four twos, then the threes... Would we not consider this arrangement also to be ordered? There are innumerable sequences that might similarly be considered to be ordered. If we shuffle the pack thoroughly the chance of any particular one of these sequences turning up is, of course, the same. Exactly the same probability, in fact, of any sequence that we might intuitively consider to be random. Furthermore, a little green man from the planet Throgmorton might recognize with delight that strange sequence as precisely the one that is used by the folks back home to start the game of Baggerooley! Extrinsic order is truly in the eye of the beholder. This kind of order merely arises from some set of humanly originated rules. A quirk of contingent spin-off. It has no basis in natural phenomena. Extrinsic order, however, is not merely characteristic of trivia. It is also found in mathematics. There is no physical ordered reality corresponding to the concept of a number-line, for instance. Or to any kind of progression. Or to fractals. That these concepts, like the other simplifications of mathematics, are enormously useful (and, for some folk, hugely fascinating) does not negate the fact that such concepts of order are of human origin.

So where in the physical world do we find intrinsic order? The short answer to this would seem to be shape. Let's first look at the common states of matter. The atoms or molecules of a gas, unless trapped by some kind of container, are free to go any which way until colliding elastically with another. We would be hard pressed to ascribe order at small scales to such a system. And we find that the shape of the gas cloud is correspondingly ill-defined and also inhomogeneous. The atoms at the centre of the cloud having shorter mean free paths than those at the periphery. In a liquid, the atoms or molecules are loosely held together by various kinds of weak intermolecular bonds such that translational movements can occur, but escape into the gas phase can only occur at any surface that is not in contact with a solid (or immiscible liquid). And then only by individuals which happen to have sufficient energy to break free. But surely, some might cry, a liquid has no shape? Oh yes it does! Some may remember Douglas Adams' little puddle which awoke one morning and marvelled at the way in which the ground beneath it was exactly tailored to fit its form.

Unlike gases, all liquids at rest have a well defined shape. Usually the shape imparted by the container. The beaker, the teacup, the lake or the seabed. On some occasions, when not contained by receptacles, their shape is determined by a balance of forces. Internal (surface tension), gravitational and aerodynamic, for instance. As in the case of a raindrop. Shape is a measure of the level of order of an assemblage of atoms or molecules. Even a flowing body of water such as a river is not entirely without shape. Accordingly, at some level, it has a certain degree of order. And flow, by its very nature, implies directionality.

With a further increase in order of a liquid at rest, translational movement decreases until, in the case of a pure substance, a rapid transition to a highly ordered crystalline state occurs. It freezes, and the heat that was hitherto represented by translational movements is discarded. For an impure substance or a polymer, transition from the less ordered liquid state to a more ordered solid state takes place more slowly, as does the reliance on a container to determine shape. So we see that intrinsic order is physically very closely related to shape. But that is not all, because there is yet another fundamental concept that we can, perhaps surprisingly, equate to shape. And that is information, the subject of a later chapter.

In the discussion of intrinsic order above, I have mentioned only the geometry of translation movements. In the case of molecules there are dynamic vibrational and rotational changes in geometry to be considered in the determination of intrinsic order, or its inverse. Each of these alternative shape variations is usually referred to as "a degree of freedom" There is another salient aspect of geometry which does not so obviously fit within the usual intuitive notion of shape. That is the related concept of a field. Of which concentration is an important measure of intrinsic order. For a homogeneous field such as a contained gas or a solution, intrinsic order (or in thermodynamic jargon, dysentropy) is directly proportional to concentration at a given temperature. Temperature itself, being a reflection of the average kinetic energy of molecules is, of course, also fundamentally an (inverse) measure of order. It, too, being reducible in terms of a vectorial field. When we come to look in a holistic way at the extreme intricacy of the great dynamic biospheric network that is biology, our intuitive comprehension of shape is near its limits. Although shape, in its usual everyday sense, is found at all scales in biology, a change of perspective is sometimes helpful. Thinking instead in terms of topologies, fields, vectors and information flows can help to make the interpretation of these a little more tractable. We can conveniently lump them under the term "organization".

Geometry, even in its most easily recognizable form, that of shape, largely defines all phenotypes. Right from the blue whale, down through the giant sequoia, the goat, the marigold, the flea, the paramecium, the coliform bacterium. Not to mention the genome.

# 6

# BIOLOGY

*A very noticeable characteristic of biological systems is the gathering and processing of information.*

*All known organisms have sensors and effectors. This is even marginally true of the viroids, for instance, which are claimed to have been produced in a back-eddy of retrograde evolution. Crystallisable assemblages that straddle the imaginary line that separates "ordinary" chemistry from that of biology*

*For the empiricist there is absolutely no good reason to equate biology with "life". Certainly it cannot be exclusively identified by replication with a degree of imperfect copying that provides variation and subsequent filtering of these variations by selection against the prevailing background.*

*Because these are also intrinsic features of technological progression. Wherein the ebb, flow and cross fertilization and simmering of ideas within collective human imagination brings forth a constant stream of new species of idea. Always fuelled, of course, by the constant influx of new data from the environment.*

*Some ideas, unsuitable for inclusion in the prevailing technological environment, fall by the wayside. Not all of these are lost forever, but, in spore-like fashion, remain dormant. Perhaps to be resuscitated at some later time when conditions have changed.*

*Other ideas and skills lead to new technologies which enable the harvesting of entirely new kinds of data from the external world. Initiating powerful feedback loops which culminate in the emergence of "inventions". Novel devices, systems and structures for processing and communicating ideas.*

*One such exploratory venture involved the recent large-scale development of rocket propulsion. Primarily motivated, as is so often the case, by warfare. Idle curiosity is seldom as well-funded! The moon landing and subsequent visits to planets have given some interesting insights. But the most significant advance, in terms of the technosphere, which arose therefrom has been the establishment of a plethora of communications satellites.*

*Another technology, again to a large extent made possible from military funds, has allowed us to extensively explore the submarine environment. This, too, has enabled the harvesting of some totally unexpected information to feed the ever-burgeoning body of science and speculation.*

*Amongst this new knowledge gleaned from the depths are findings that, when combined with our current wisdom of chemistry, at last provide a very plausible mechanism for the advent of the particular phase of chemical auto-synthesis which we know as biology.*

*This pathway to abiogenesis is addressed in chapter 9*

# 7

# PLANET EARTH

A comprehensive assessment of the generation of chemical intricacy of the outer layers of our home planet has been put forward by the geologist Robert M Hazen. Although couched in evolutionary terms which make some of those who remain locked into traditional mind-sets nervous, his findings seem to be generally well accepted by his peers. I outlined these findings in a previous work but they are well worth repeating here:

The precursors of solar systems such as ours comprise a central mass surrounded by an orbital disk of dust and gas. Hazen estimates the minerals in this disc to be limited to only about a dozen types. When the density of the central mass becomes sufficient to allow fusion, the star then blazes forth with heat pulses which increase the mineral diversity of this early accretion disc. Further thermal and pressure variations that result from the progressive condensation of the original dust into quite large planetesimals then bring the mineral inventory up to around 250. Increased, yet again, by the reheating of volcanic material to produce granites, a grand total of, maybe, 500.

The presently favoured model of the moon's formation, that of being side-swiped by a hypothesized body of about the size of Mars would have resulted in a great dispersion and degradation of the mineral species which had arisen up to that stage. But a new evolutionary process was to emerge. As the earth cooled, solid crust formed on its surface, forming the hypothesised single land mass of Rodinia.

Convection currents in the molten interior and the continual formation of new crust eventually giving rise to subduction and the fragmentation of Rodinia, the advent of plate tectonics and a continual swallowing up of broken parts of the Earth's crust. This repeated thermal cycling driving the concentration of mineral species in a manner somewhat reminiscent of the zone refining technique used in industry today. In addition, the further evolution of yet more mineral species, which Hazen estimates at being of the order of 1500. And there it would have stayed if not for the novel phenomenon of biology. A new kind of chemical intricacy which had been quietly ticking away in the background for a billion years and which was to provide the new prevailing conditions for this next increase of intricacy to pick up speed.

This bounty provided by biology was, of course, the generation of atmospheric oxygen. As a result of which the evolution of yet another 2,500 "naturally occurring" terrestrial minerals has occurred by present times. Many of these, in turn, repaying the debt owed to concurrent biology by providing the correct prevailing environment for some of its own subsequent evolutionary steps. Not to mention the investment in preparing the substrate that would much later, with the advent of human imagination, not only enable but also make inevitable the evolution of technology.

Robert Hazen is clearly very conscious of the kind of model advanced herein and perceptively senses the continuity of nature's evolutionary machinery. At least up to the stage where of our kind of imagination arose.. He remarks:

> *"In this sweeping scenario, minerals represent but one inexorable step in the evolution of a cosmos that is beginning to know itself."*

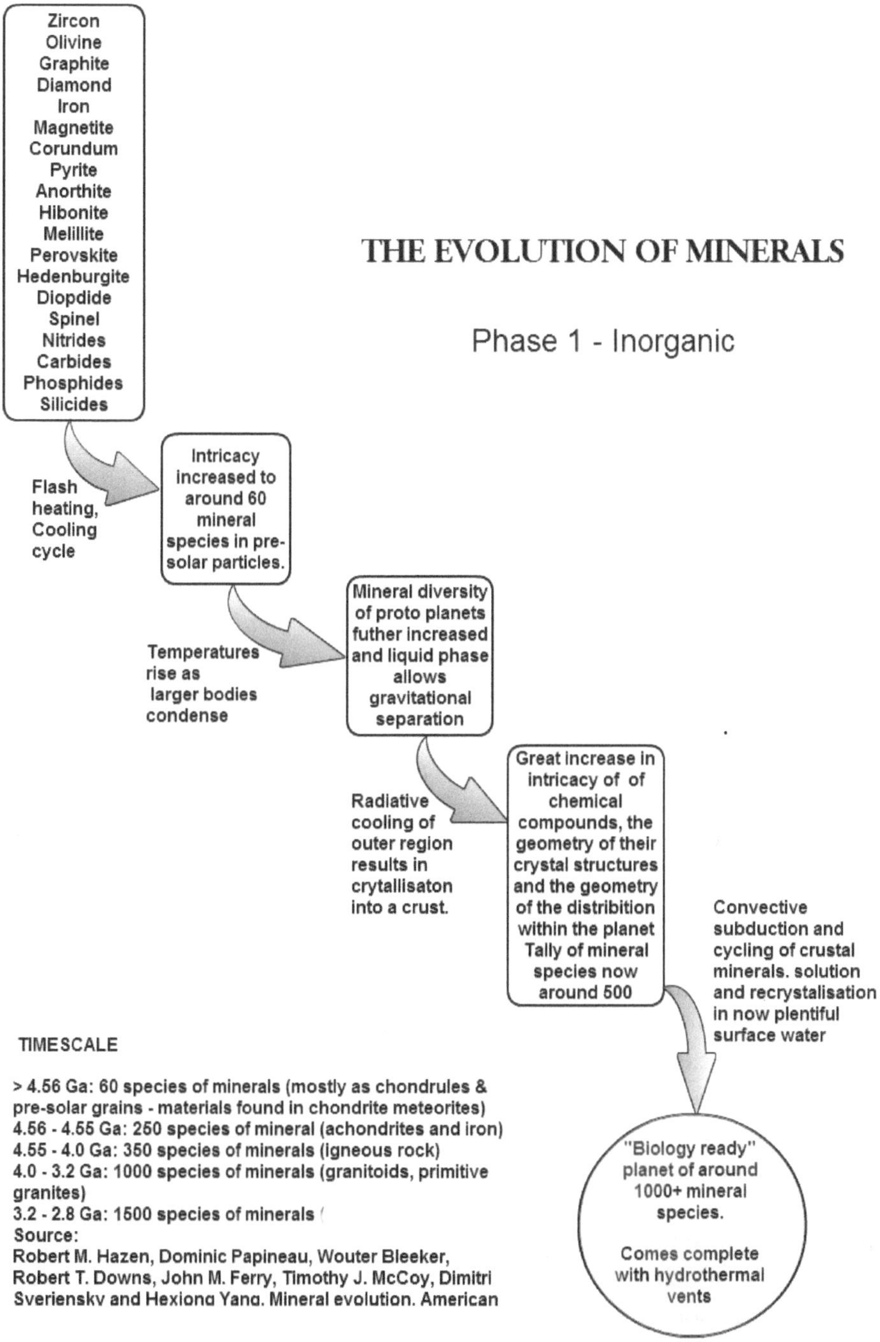
THE EVOLUTION OF MINERALS
Phase 1 - Inorganic
Zircon
Olivine
Graphite
Diamond
Iron
Magnetite
Corundum
Pyrite
Anorthite
Hibonite
Melillite
Perovskite
Hedenburgite
Diopdide
Spinel
Nitrides
Carbides
Phosphides
Silicides
Flash heating, Cooling cycle
Intricacy increased to around 60 mineral species in pre-solar particles.
Temperatures rise as larger bodies condense
Mineral diversity of proto planets futher increased and liquid phase allows gravitational separation
Radiative cooling of outer region results in crytallisaton into a crust.
Great increase in intricacy of of chemical compounds, the geometry of their crystal structures and the geometry of the distribition within the planet Tally of mineral species now around 500
Convective subduction and cycling of crustal minerals. solution and recrystalisation in now plentiful surface water
"Biology ready" planet of around 1000+ mineral species.
Comes complete with hydrothermal vents
TIMESCALE
> 4.56 Ga: 60 species of minerals (mostly as chondrules & pre-solar grains - materials found in chondrite meteorites)
4.56 - 4.55 Ga: 250 species of mineral (achondrites and iron)
4.55 - 4.0 Ga: 350 species of minerals (igneous rock)
4.0 - 3.2 Ga: 1000 species of minerals (granitoids, primitive granites)
3.2 - 2.8 Ga: 1500 species of minerals
Source:
Robert M. Hazen, Dominic Papineau, Wouter Bleeker, Robert T. Downs, John M. Ferry, Timothy J. McCoy, Dimitri Sverienskv and Hexiona Yana. Mineral evolution. American

In a review of a paper in "American Mineralogist, Minik Rosing adds:

> *"As Hazen et al. suggest, this cause of mineral species diversification must be a general feature of Earth-like planets. On Earth itself, the emergence of life gave rise to new opportunities in the mineral world. Living organisms can construct chemical compounds that are not favoured by thermodynamic equilibrium in the ambient environment and that would not have formed spontaneously without the intervention of life. These compounds (for example aragonite in some animal skeletons) provide advantages for an organism that make it worthwhile to expend energy acquired from other life processes to synthesize them. The intervention of life in Earth's geochemical cycles therefore greatly expanded the range of mineral species in Earth's surface environments, and added minerals such as the sulphate mineral gypsum that we use in drywall panels and the haematite used in red paint. This process coupled mineral evolution to biological evolution in that some of the diversification of mineral species was caused by diversification of biological species."*

There is, of course, a further and much greater proliferation of chemical intricacy which extends far beyond the domain of geology. It would seem reasonable to include this multitude of novel inorganic compounds, alloys and crystalline forms which, in recent times, have emerged from the machinations of human imagination. While no match for the almost infinite potential for variety that is inherent in the bond structure of the carbon atom, which has generated intricacy beyond compare, these new inorganic entities, millions of which have already been identified, dwarf the pre-human 4500 or so "naturally occurring" minerals. I use the quotes advisedly because it is fundamental to the theme of this book that all is the product of nature's evolutionary machinery. It is a nonsense to arbitrarily partition our species and its artefacts off from other natural processes. But old anthropocentric habits die hard!

## THE EVOLUTION OF MINERALS

### Phase 2 - Organic

A MAJOR BREAK-THROUGH IN THE INTRICACY OF CHEMISTRY AND GEOMETRY

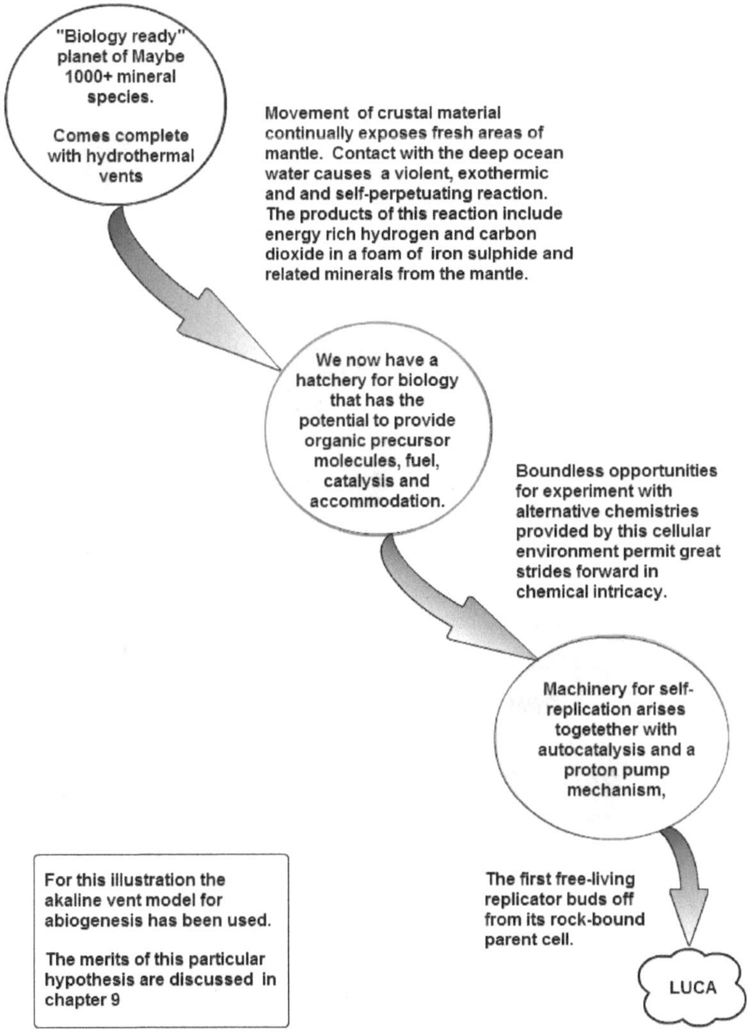

# 8

# GAIA

The Gaia hypothesis put forward by Lovelocke and Margulis in the latter part of the last century got off to a bad start.

The choice of name alone, that of a Mythical goddess, was unfortunate. The taint of some vaguely teleological connotations played upon the irrational fears of certain scientists, particularly among those more prominent ones overly concerned with maintaining respectability within their peer group.

Furthermore, these same properties provided excellent hooks to which the superstitious and the mystically inclined could latch on. The unjustifiable pseudo-scientific nonsense that they spawned further muddying the scientific waters. Which is sad, because the underlying notion is simply a statement of the very obvious fact that all organisms, as well as many aspects of Earth and its atmosphere are, at some level, interrelated and interactive.

The undeniable proposition they are all part of an intricate network that spans the globe. A dynamic network in which feed-back is found to be ubiquitous. Both negative, promoting homoeostasis or even inhibition, and positive, kindling amplification or rapid change. In this form, at least, Gaia does not even justify description as a hypothesis. Rather, it is simply a very well founded observation.

Another feature of the original version that inhibited acceptance within scientific circles was the undue emphasis on cooperation, while understating the important role played by competition. (Actually, as I shall argue later, viewed from the perspective of the network, rather than that of individual organisms, the distinction between cooperation and competition blurs)

And it is from the network, the biosphere in its entirety rather than from its components, that the observed directionality emerges. That the ratcheting forward with respect to time occurs. The observation that it is difficult to ascribe this, at a global level, to natural selection is entirely valid but of little relevance. As far as we can tell, selection seems to be an identifiable feature of only the biological and technological phases of the overall evolutionary process.

Thinking in terms of networks can also clarify many issues at far smaller scales. The organisation and structure of swarms. For example, of ants or bees. Or of fish and birds that exhibit collective behaviours. Or, for that matter, of individual organisms. We ourselves, are in some contexts more usefully thought of in terms of a highly integrated swarm of cells. A cooperative networked society. Chapter 11 pursues this theme.

It is even appropriate for the individual cell, similarly an exceedingly intricate cooperative network, in which, contrary to the paradigm which is currently fashionable, DNA turns out to be a relatively small player . A cell always comes from a cell by division. Its structure and its functional components are directly inherited from its predecessor. This train of reasoning, although very solidly founded, leads to way of viewing the interplay between the genome and the cell in a very different perspective. This challenge to orthodoxy is also to be found in chapter 11

***The graphical representations on the following two pages give some small indication of the interlinked intricacies of the biospheric network.***

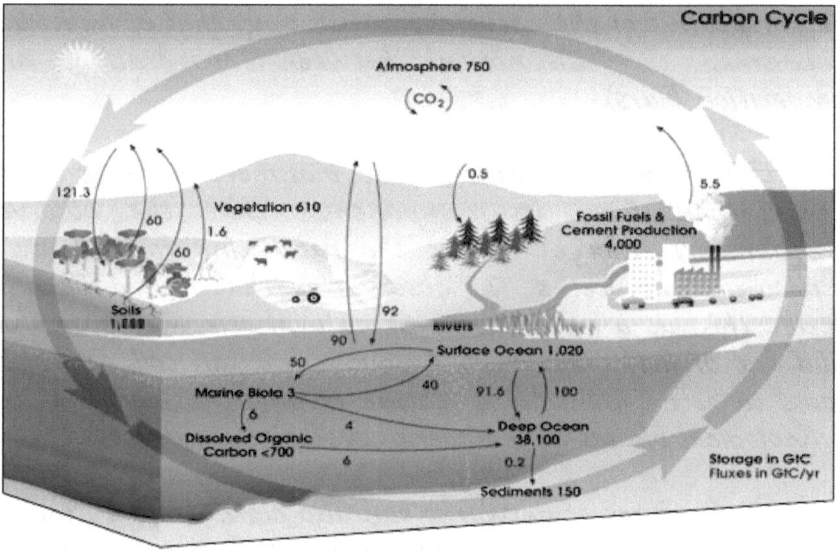


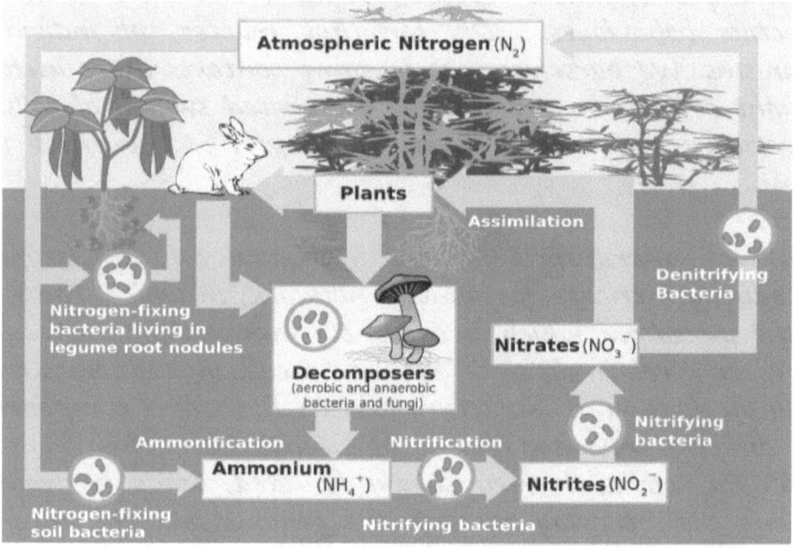

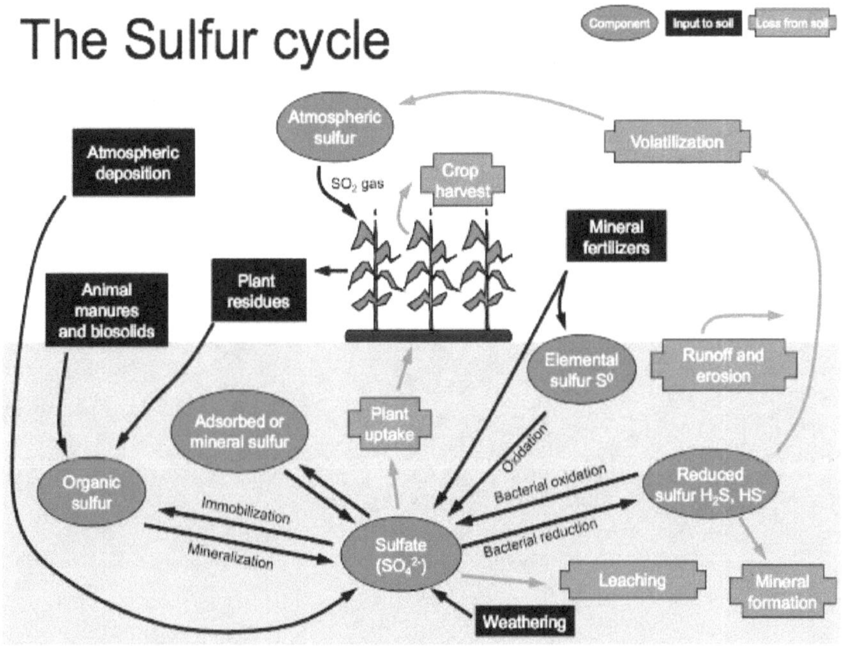
The Sulfur cycle
Component
Input to soil
Loss from soil
Atmospheric sulfur
Volatilization
Atmospheric deposition
SO2 gas
Crop harvest
Mineral fertilizers
Animal manures and biosolids
Plant residues
Elemental sulfur S0
Runoff and erosion
Adsorbed or mineral sulfur
Plant uptake
Oxidation
Organic sulfur
Bacterial oxidation
Reduced sulfur H2S, HS-
Immobilization
Mineralization
Sulfate (SO42-)
Bacterial reduction
Leaching
Mineral formation
Weathering

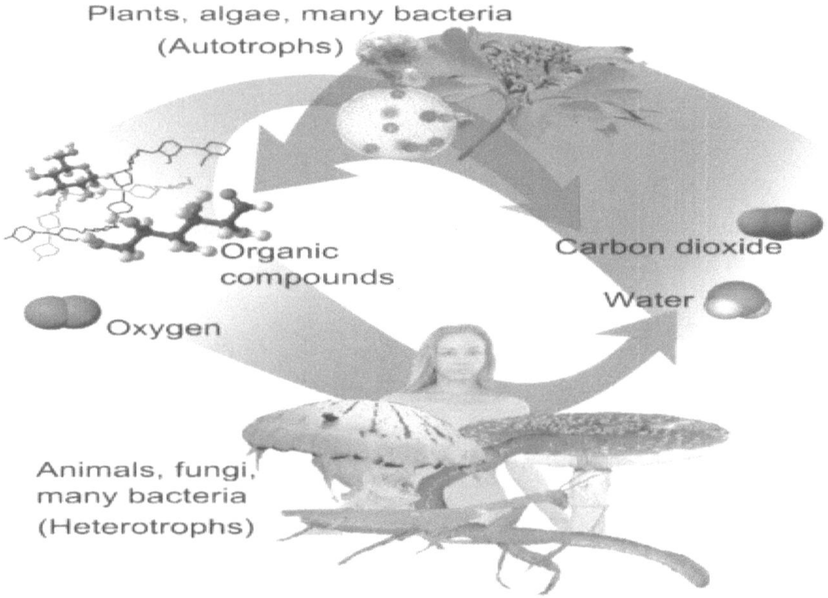
Plants, algae, many bacteria
(Autotrophs)
Organic compounds
Carbon dioxide
Water
Oxygen
Animals, fungi, many bacteria
(Heterotrophs)

# 9

# ABIOGENESIS

Our exploration of the concept of an evolutionary continuum renders the otherwise vexing question of the "beginning of life" meaningless. Simply because, using our wider interpretation of the "life" process, there seems to be no sensible reason for ascribing a "beginning" to any particular part of it. Certainly there are identifiable transitions between the various phases. But each of these emerges as an immediate consequence of that which precedes it. The "Big Bang" scenario, the presently accepted best guess, made on the basis of such evidence as is available to us, describes an evolving universe from as far back as that singularity and can say nothing about events, if there were any, preceding that "time". However, since all we observe in nature is an evolutionary continuum, a continual increase in intricacy, the speculation that is most consistent with these observations is that of a "budding off" from a parent entity. It is a heuristically compatible speculation. It also has some very tentative observational support arising from studies of cosmic microwave background (CMB) by Sean Carroll & Adrienne Erickcek's Caltech team. Using data acquired from NASA's Wilkinson Microwave Anisotropy Probe (WMAP) they claim to have gleaned a hint of a parental origin from this data. So perhaps the "Big Bang", a term first used derisively by Fred Hoyle, a champion of the steady state theory of "continuous creation" which had gained widespread acceptance before the serendipitous discovery of the CMB, might be better served by "The Big Bubble"

While there is no direct evidence for (or against) this idea we have absolutely no good reason to suppose that the overall evolutionary pattern we see in nature's machinery should be assumed to happen to break down at this point. Heuristically, it is a sensible default assumption that it does not. Which is about the best we can do for a domain that is not directly accessible to science. Not to be forgotten, of course, is the only other gross phenomenon observed in our region of space and time. The seemingly inseparable correlate of this broader evolution, entropy. Indeed, the overall increase in entropy over time can perhaps be usefully regarded as the stochastic cost of intricacy.

Rather than engaging in the futile quest for "beginnings", we are far better served by trying to identify the periods of transition between the various (often overlapping) evolutionary phases. As in that shift which is usually referred to as "abiogenesis". The emergence of the particular kind of replicative chemistry and geometry that we know as biology. A common characteristic of biology is positive feedback. Indeed, it is this which accounts for the amplification which has occurred since the emergence of the prototype cells to give the vast biomass we see today. Autocatalysis, the ability of a molecular species to utilize an energy source to generate a similar molecule, is perhaps the simplest form of chemical replication. More common are autocatalytic systems, wherein a multiplicity of species and/or pathways are involved. Catalysis can be defined as a substance providing a geometry such that the probability of individual molecular interactions are increased, thus increasing the overall rate of reaction. For example, in the case of a metal like platinum, adsorbed species are constrained on the surface such that reaction occurs more readily than by random collisions in the bulk gas or liquid phase. Similarly, in biological systems the geometric configuration of enzymic proteins specifically align the substrate molecule(s) to facilitate the required outcome. The catalyst is not consumed by the reaction, nor does it , of itself, determine the direction of the reaction.

However, we can attempt to establish a property, or set of properties that define biology. One very fundamental way of implementing this would be to equate biology with "The first age of electronics".

This works quite well because biosynthesis and all other work done by the cell is fuelled by the shuttling of electrons from a high energy source to a lower energy sink by molecular transfer. Quite distinct from "The second age of electronics" which first really got into its stride in the early nineteenth century and instead utilises the flow of free electrons through conductors as an alternative way of both performing work and increasing intricacy. Again, we might choose to use information processing as the defining characteristic of biology. Except that, although less easily recognizable as such, all chemistry can be argued to involve information processing. And, of course, today we have non-biological computers dedicated to such tasks. So, while applicable, this definition lacks exclusivity.

The definitions most popularly accepted in recent times involve replication of chemical species, or assemblages thereof, with the potential for accidents, copying errors or shuffling that allow of selection. In other words, filtering by the prevailing environment of stochastic variations.. Although some experimental attempts are being made to synthesize alternative systems which could meet these requirements, none have been so far successful and thus these rather simplistic criteria for biology seem to hold up quite well as a rule of thumb. With one confounding exception. That being, once again, the evolution of ideas within the collective imagination of our species. Which similarly proceeds by virtue of random inputs, replication, copying errors, shuffling and melding. Again, filtered by selective processes. In this case, of ideas. An issue to which we will consider rather extensively later on. The particular chemical and geometric pathways leading up to entities meeting even these rough and ready biological criteria, however, are exceedingly complex, speculative and subject to much contention. While there is little doubt that a plethora of organic moieties providing the potential building blocks for biosynthesis were to be found in both the oceanic "primordial soup" and the atmosphere, the requirement for concentrations sufficient to allow high degrees of interaction is problematic. As is the provision of catalysis and a suitable energy gradient. Another question is whether the components comprising the cell (for, as we shall see later, all biology, including even the supposedly retrograde viruses and viroids, is based upon the cell) actually co-evolved, evolved separately and subsequently merged, or evolved serially The most clearly identifiable components required by every known ***existing*** organism are:

1. A dedicated data source oligomer on which the recipes for generating various particular sets of proteins are digitally encoded. DNA is the only known example that has been observed in extant organisms. Some of the more recent research suggests that alternative nucleotides and oligomer backbones might allow their substitution, but whether their properties are sufficient as to be able to function within an aqueous cell environment is, as yet, very questionable. It is further questioned if the specialized databases represented by such elaborate oligomers were a feature of the earliest replicators or evolved later.

2. The machinery for generating proteins from amino acids using recipes read from the DNA data source. In extant organisms this corresponds to the host of intracellular enzymes which generate, process and transport the RNA variants for guiding the assembly of proteins by ribosomes. Again, in the event that nucleic acids were not originally used for this information storage, the feature may not have been required.

3. The cell membrane. Overly simplistic models have been proposed for this which are based upon the quite common formation of lipid vesicles under a variety of conditions and are widely touted. I must admit to having been guilty of this myself having, in "Unusual Perspectives" spoken of:

> *"one of these replicators ... would inevitably, some day, wind up sitting inside one of the oily blobs commonly formed by these hydrophobic compounds. With this lipid or coacervate jacket, this little fatty suit of armour, inside which there might, too, be room to store a few provisions, survival chances were suddenly rather bright. By some process akin to this the cell was born."*

However, this very naïve model really does not work. The cell does not, and must not, isolate the internal system from its environment. It is absolutely essential that nutrients are taken in and that waste products are excreted. The mechanisms of the cell which enable these transfers are very sophisticated and it is almost inconceivable that a primitive lipid vesicle, even if perforated, as in some experiments with cell lysates, could actually form so as meet these requirements. It is sometimes argued that virus or the viroid are examples of a cell not being a prerequisite for biology. This, of course, is very wrong. Neither of these can function without using a cell as a necessary part of its parasitic life cycle. There are no observed examples of acellular biology.

4. Proton pump mechanism to provide a reservoir of energy derived from electron transport chains. This is universal in all presently known organisms.

Another enigma of the cell is that its boundary appears not to be coded for by data source oligomers or manufactured by the ribosome. The information for its chemistry and geometry seemingly being derived directly from that of its predecessor. A cell always arises from the division of a cell. And, as far as we know, always has done. As exhaustively underlined by Franklin M Harold in his excellent book "The Way of the Cell".

So it is interesting to speculate that maybe the cell arose independently and in some way evolved before the incorporation of the data source and protein assembly systems that it contains. The problem with this is that although the cell boundary is a phospholipid bilayer, the highly selective ports which enable interaction with the external world are proteinaceous. Without these the cell is not viable and hence co-evolution, at least with proteins and some kind of data storage is necessarily implied. How could such at thing come about? We will return to this question later, but will first take a broad look at the merits and demerits of some of the proposed abiogenesis pathways.

Creationists point to the chances of any such concurrencies occurring as enormously improbable. and they are right. But there is, of course, a way round this. The chances of your ticket winning the national lottery are extremely low. But if you buy a million tickets they suddenly become rather good. and if you buy a million tickets, week after week, your win becomes almost certain. Firstly, to consider the panspermia hypothesis. That the seeds of biology are distributed throughout space and pretty well "hit the ground running". This is not widely accepted in scientific circles for much of the same reasons that it does not tie in with the evolutionary "big picture" we are developing here. But the biggest objection is that it does not address the issue of how these proto-biological entities were formed in the first place.

The popular and very naive "primordial soup" model of abiogenesis occurring in the open ocean suffers from the problem of dilution. Preventing sufficiently close interaction between molecular species to be stochastically favourable. This is also even more so in the case in the atmosphere. The lack of a sufficiently high and consistent energy gradient is a further problem. Ultraviolet radiation, lightning and volcanism being the main contenders offered. The ready hydrolysis of nucleotide precursors is another issue. Yet another serious problem with the open-ocean hypotheses is the presumed extremely low concentration of phosphate, which modern biology requires both for the high energy bonds required for biosynthesis and other processes, as well as for incorporation in nucleotides .

The concentration problem can be mitigated in the case of partially dried ponds and, as well, by adsorption of appropriate moieties on mineral surfaces such as clays, which could conceivably also guide and catalyse reactions at lower energy levels. Porous materials, by virtue of their very high surface area have great potential in this respect. Even gases and other volatile compounds can be trapped and thus concentrated by adsorption. And, as a bonus, hydrolysis of precursors could be held at bay. This then, at first glance, would seem to give encouragement to those who choose to hypothesise a partially atmospheric generation of sufficient useful molecular building blocks. Unfortunately, there are flaws in this idea:

- Firstly, the supply (and sink) of electrons. For even the organic molecules of pre-biology are dependent on this for any real increase in intricacy. The Earth's surface is bathed in photons that can generate high potential energy electrons. A great abundance even now, and far more in those ozone-free days. But sunlight will not penetrate into the micro-labyrinths of porous materials. Nor, ordinarily, will lightning. Now, volcanoes do produce some porous rocks and also can, for a while at least, provide a supply of electrons. So are volcanoes, which also have their abiogenetic theorist fan club, really a good candidate? Again, sadly, no.

- Because of the second flaw: For although porous (or finely particulate) materials are very good at adsorbing molecules, and because of this, also have a potential for catalysis, once all the adsorption sites are occupied and any interactions have taken place, the system will become static, will reach equilibrium. There is really only way out of this dilemma and that is flow. Flow though a porous or perhaps finely particulate material gives the possibility of continuous dynamic change, of increase in molecular intricacy. Also of carrying high energy electrons to active sites. It even provides for a particular kind of selection. Selective adsorption efficiently enabling the separation of molecular species. This is the basis of the various kinds of chromatography system to be found in any chemistry lab: A flow of gas or liquid through a particulate/porous medium.

It is hard to see where flow could originate in the solidified rocks of volcanoes. Perhaps in intermittent spewings of gases or in heat cycling? Similarly with the evaporated pond-side. The oceanic tidal zone could perhaps generate a cyclic flow, the pores being filled at high tides and emptied at low. Or a river might provide a way out, but whence the abundant supply of electrons? We seem to be scraping the bottom of the barrel with all these scenarios. None of them really seem to fit within the established principles of physics and chemistry.

Even if we resort to the panspermia model, the imported proto-life "seed" would seem to have a hard time trying to survive and prosper. However, it certainly has to be admitted that, despite their inherent weaknesses, these ideas ***all*** give far more plausible pictures than the old guy in the sky making it all out of plasticine, or Maui pulling the islands of New Zealand (where I happen to live) out of the sea on his fishing line, or bizarre concatenations of events involving elephants.

In recent years, interest has been gathering in deep ocean vent processes, where many new organisms have been discovered. The vents provide very favourable energy gradients as well as a good concentrated variety of possible biology precursor molecules. They fall into two classes, the acidic "black smoker" volcanoes, releasing hydrogen sulphide, and also the alkaline hydrothermal vents, which, according to some (not all) have their origin in the tidal forces of the .Earth-moon system cracking open the sub-sea strata

*Mid-Atlantic alkaline hydrothermal vent structure.*

There are advocates for either of these as sites for abiogenesis. However, the alkaline vents have the significant advantage that, as well as supplying many likely precursor building blocks in addition to concentration by adsorption, catalysis, flow and energy gradient, the porous structures generated by their activity also provide what can be considered to be possible moulds or matrices for the evolution of the primitive cell. Here we can find, at last, a very plausible scenario wherein the precursor compounds such as thiol esters might be compartmentalized within the microscopic cavities in the structures that arose from the activity of these vents. As Ephraim Racker points out:

> *"thiol esters may possibly represent the evolutionary ancestors of 'high energy' ATP"*

Also, even the possibility of gleaning the intractable rock-bound phosphate ion from its very low solubility calcium compound.

To help us bring the situation into focus, let us consider the analogy of a vast organic synthesis laboratory with a billion flasks, rather than the one reaction vessel used by Miller and Urey. Each flask, it is very reasonably hypothesized, lined, in those anoxic times, with a catalytic surface of iron pyrites. A flowing conveyor system loads each of the flasks with a random blend of molecular species of the kind which represent likely building blocks of primitive biology. The flow also inducing turbulence within the flasks. An energy gradient resulting from the reaction between the abundant acetyl thioester and carbon dioxide (or monoxide) being provided to augment catalysis in driving whatever increases in molecular intricacy and diversity may occur. Furthermore, the warm, reductant efflux is alkaline causing a proton gradient between it and the acidic ocean.

In this way we have billions of experiments being performed on a stochastic basis in a reticulated open cell foam, with all manner of syntheses occurring simultaneously. Nature's machinery appears to have scant regard for economy, and the billion reaction vessels of our analogy are most likely on a small scale compared with the experiments that could have been similarly performed in the myriad interconnected cavities of the hydrothermal vent structures over fairly long periods.

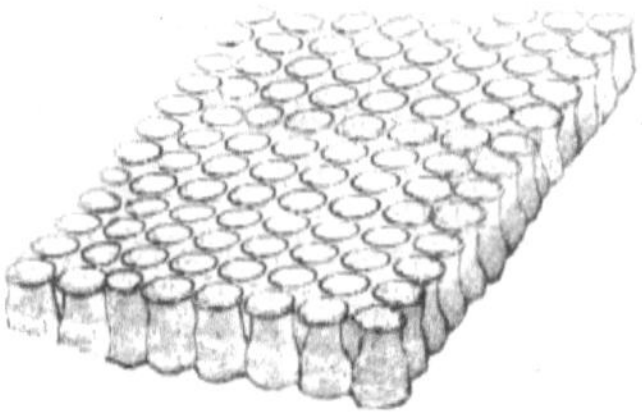

Portion of "zettawell" plate

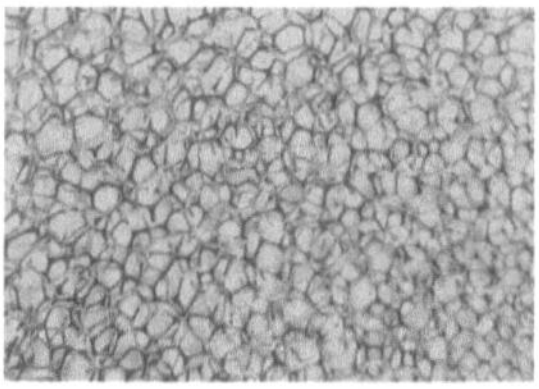

Reticulated foam

Here we find solved the problems that plague the oceanic, land-based and atmospheric models. We also find a basis for the cell itself, inasmuch as we might reasonably expect the deposition of lipids on the surfaces of some of the cavities. While the probability of interaction is enhanced by containment, these protocells are, at this stage still open to ingress of nutrients and egress of waste so these are not issues. Furthermore, because we have flows that hold chemical equilibrium at bay, some degree of "spillover" can be expected between cavities that provide transfer of evolved molecular species. Thus increasing the overall diversity of the system.

At some limiting stage, however, the cost of disruption to a well-functioning system of this kind might be expected to outweigh any advantages incurred by such a transfer. In which case a "selection pressure" (of differential survival rather than differential replication) would thus arise, favouring some kind of selective enclosure for such enclosures to progressively evolve. The end-point of this process resulting in a functional membrane with mechanisms selectively isolating the machinery within its cavity from the external environment. A primitive form of the barrier found in the membrane of the cell.

We can fairly safely speculate that in the most "successful" prototypes, alternative autocatalytic processes, including that of the provision of a proton gradient, may have arisen. Developments that reduced reliance on the metallic sulphide lining of the cavity.

We can also posit a selection pressure for evolution of the closure membrane to bulge outwards as a bubble. thus increasing its surface area and hence its access to the original or alternative fuels and electron acceptors. The logical final outcome to this trend being for this bubble section to maximize its surface area within the constraints of surface tension and to then break away. Complete with internal autocatalysis and selective input/output mechanism including a primitive implementation of the proton pump. So, here too, we have a very plausible mechanism for the inception of cell division.

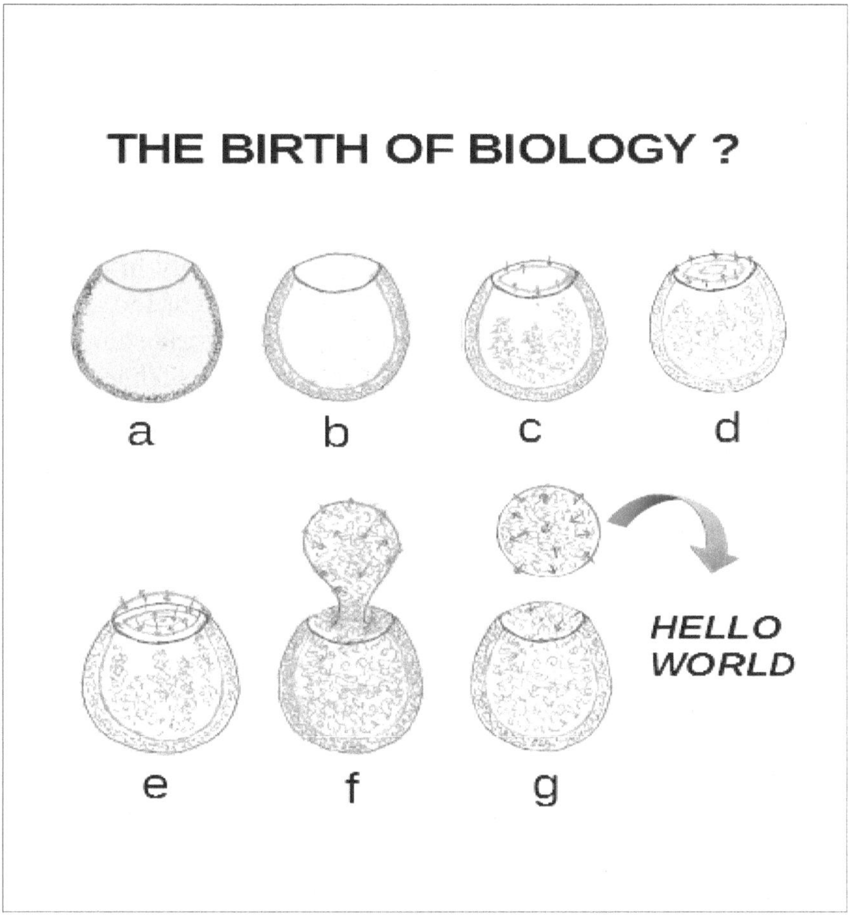

Having, in this way, acquired a relatively "free-living" lifestyle, these hypothetical primitive cell(s) would still, of course, be way short of charging boldly forth to make their livings in the big wide world of the oceans. They would be constrained to the nursery environments until such time that further adaptations, such as, perhaps, a cell wall and modifications of internal chemistry so as to employ alternative electron resources, allowed a gradual venturing further afield. Thereby setting the scene for much greater chemical and geometric intricacies. This model, so far, can explain the production of basic cells that are no longer tied directly to the compartments of the inorganic matrix in which they had evolved. Since concentration, but now in terms of protocell population, is still an issue, at this stage we have to envisage these primitives somehow "tied to the vents" for some time and further evolving by interchange of components and mergers before venturing further afield. Those who prematurely "float off into the ocean" never to be seen again. Those that stick around and mingle until fully equipped for the freedom and perils of the big wide world faring better.

It could be asked why we don't see this kind of activity in the vents today? The answer to that is simple. Partly because of the action of biology itself, the window of opportunity has passed, the composition of the Earth and its oceans have changed in ways that such events can no longer occur. In particular, they have become oxygenated. Even if this were not the case and some similar kind of process could still take place, this deep sea environment is today swarming with predators ready to swiftly gobble up any fragments of proto-life. Indeed, the recent surprising discovery by Ben Thuy et al of fossils dating back 180 million years suggests that early biology may have lingered in the ocean depths far longer than previously imagined.

With the advancement of a cell structure with internal autocatalytic systems and closely related extra components we have an entity from which the biological systems of today could have evolved by natural selection. In terms of probability, the assembly of nucleotides and incorporation into the database oligomer seems to be the main stumbling block. But this is ameliorated somewhat if we (very reasonably) assume that adenosine triphosphate (ATP) evolved first for powering the cell and was later exapted, along with variants, for information storage. We have a situation that can be likened to billions of monkeys generating the works of Shakespeare by randomly typing away at keyboards. But the prototype replicating and divisible cell we have here contains far less information than the writings of the bard of Avon! And its selective ratcheting (not shared by the unbiased monkey model) is effected, as always, by the underlying structure of our universe. There is no direct evidence, of course, to support this chain of events. However, it is in my view perfectly valid to draw upon the "big picture" that is provided by the sciences in their entirety to settle upon a provisional "best guess" in cases where a believable mechanism is to bc found. Consilience!

The scenario also raises another interesting possibility. That the precursors of the archea and the bacteria might have separate origins. These two lineages, while having some features in common, have some very significant differences. Most notably the entirely unrelated chemical structures of their respective cell membranes. The only feature that the two kinds of membranes share is that they are both phospho-lipids which, given the available starting materials, may well have been the only option to provide a barrier comprised of units that are hydrophobic at one end of the molecule and hydrophilic at the other. There any similarity ends. The structures could hardly be more different.. The hydrophobic portion, the "tail" of the bacterial cell molecule is a fatty chain of around 16-18 carbon atoms. Whereas the kindred "tail" of the archeal equivalent is an isoprenoid chain of about 20 carbons with side-branches . This is joined to the hydrophilic end of the chain with an ester bond for the bacterium, whereas an ether bond is used for the archean. But even more telling is the fact that the asymmetry of the glyceryl phosphate moiety makes the molecules chiral. In bacteria the ***d*** isomer is used, in bacteria, the ***l*** isomer. There is an incompatibility between right handed and left handed molecules that results in chirality being conserved in biological systems. This in itself, strongly suggestive of separate origins.

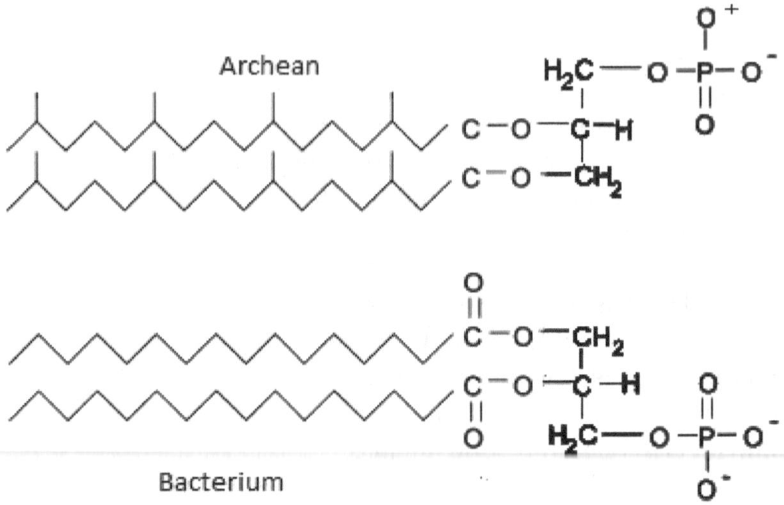


And, after all, the genome does not appear to directly code for the basic cell membrane, only for its proteinaceous inserts. A cell only ever arises from the division of a cell. Inheriting the membrane and a complete set of all the internal components, including ribosomes, which will be absolutely essential for its subsequent interaction with the nucleic acid database so as be able to read the instructions for manufacture of the more of the proteins it needs for growth. The features that bacteria and archeans ***do*** share may have arisen independently as close neighbours in the original fixed protocellular community. Perhaps among other kinds who did not survive their newly found freedom. As Carl Woese has pointed out, because of the effects of horizontal gene transfer, no individual can properly be considered to be a last universal common ancestor. The lineage being necessarily diffuse. So, for biology as a whole, the concept of a last common ancestor (LUCA) would, in this sense, thus seem to be fundamentally flawed. At root, maybe we are all mongrels.

A strong feature of this alkaline vents model is that it provides a robust riposte to creationists and the like who are apt to aver that there is no way that science can explain how (biological) life could have originated. This ***is*** a quite sound explanation of how it ***could*** originate.

Whether it is the right one or not still remains to be seen. Perhaps!...

This model, loosely based upon that promoted by Martin and Russell, is certainly not without controversy. Jeffrey Bada, who favours the Urey/Miller scenario tied to volcanism, points to the instability of amino acids when exposed to temperatures and pressures that are comparable to those found at hydrothermal vents. Also to the racemisation of aspartic acid as a potential problem. But a strong additional point that is in its favour is the fact that a reaction product of the acetic thioester with carbon dioxide is pyruvic acid - an important component of the citric acid cycle that is crucial to all aerobic respiration. Furthermore, the reaction with the phosphate ion produces acetyl phosphate. the presumed evolutionary precursor to the adenosine triphosphate that energizes the citric acid cycle. Christian de Duve, in "Blueprint For A Cell" speculates on various possible molecular pathways for a such a "thioester world"

Moreover, the model has the virtue of giving a basis for experimental study. The provision of hypothesized nutrient systems to a large cellular matrix for instance. Admittedly a long shot! NASA is among those working along these lines. In common with some of the seashore and volcanic models, its initial dependence upon one or more mineral substrates for catalysis is consistent with the homochirality that is observed in biological systems. As Robert Hazen points out:

*"A key attribute of life, and an important consideration in origins-of-life models, is life's molecular handedness, or chirality. Chiral crystalline surfaces provide effective environments for discrimination of left- and right-handed molecules in both natural and industrial contexts "*

While, within the context of our broader evolutionary model, we cannot justify a "beginning of life" status to biology, it is fair to say that is does at least represent a quite well defined phase transition in the relentless advance of chemical intricacy and diversity. We must further extend our definition of fully fledged biology to include a very proactive departure from thermodynamic stability. One that requires the input of materials, a processing stage, and output of wastes. For this to be efficient it is also necessary for the organism to have the capability to input information, and having processed it, to output information in the form of a response. (The "holy trinity" of matter, energy and information is discussed in chapter 7).

Biology may thus be quite well identified by the extensive presence of feed-back loops, both internal, in the guise of autocatalytic mechanisms, and external, where information in the form of chemical species and also energy are actively extracted from the environment, processed and returned. Different molecular species and low grade energy being output to the biosphere. In fact, we could quite succinctly and accurately, define biological life in terms of cell based entities having inputs, outputs and structures which are specifically dedicated to processing information. And here, even in the simplest creatures known today, even that retrograde parasite, the virus, we have the glimmerings of a kind of consciousness. Those sets of chemical and electrochemical informational feedback loops which we can identify as awareness that have, over the aeons, burgeoned throughout the biological realm. Most notably in humankind, where the edifice of sophisticated molecular species from which it has arisen has provided a platform for yet another quite distinct phase of the evolution of chemical intricacy. That is the evolution of technology within the collective imagination of our species. We will address this issue in later chapters.

While we are, as above, able to draw a plausible broad pathway for abiogenesis, one worrying aspect remains. Why, if biology arose after this fashion, should there be evidence of only one successful instance? Only one universal common ancestor? We have already looked at one way around this. The same issue arises much later in bio-evolution with the emergence of the eukaryote. The creationist argument of the improbability of such a complex entity "popping up out of the blue" by sheer chance is, again, entirely valid. The prospect of a whirlwind assembling a Boeing Dreamliner from a pile of junk is far from convincing. But that is not how we see evolutionary processes, biological or otherwise, to work. With few exceptions, we see that they occur over large concentrated populations and, although the rate of change may vary greatly, diversification takes place mostly in very tiny steps. The safest assumption, by far, is that by such incremental processes arose a new niche for the further evolution of chemical intricacy. The very prolific phase that is biology!

In the alkaline vents model we have not addressed the issue of the extent to which some of the most difficult molecules to assemble, the database nucleotides, might have evolved. Much is made of the "Central Dogma", a term first used (and later retracted) by Francis Crick. This essentially avers that information flow is always from nucleic acid to protein. Never in the reverse direction or from protein to protein. This is still popularly accepted for present day instances with the exception of prions. Prions are "miss-folded" proteins which are infectious agents, replicative only in the sense that their interaction with normally folded proteins can cause these to adopt the anomalous configuration. While a transfer of information indubitably takes place, it is more of a "domino effect" and not really comparable with that associated with inheritance.

There are, in any case, more serious flaws in the Central Dogma hypothesis. These will be discussed in the next chapter. Leaving that aside for the moment, however, Crick's proposition was intended to be applied only in respect of present organisms and has been very improperly used to imply a necessarily similar situation in pre-history. This apparently being the main reason for his subsequent retraction. Neither the "protein first", the "RNA first" nor the "DNA first" advocates have any sound basis for their beliefs. Just as weak is the "cell first" hypothesis. In fact all ingenuous hierarchical assumptions of this kind must be considered suspect. While the ease of assembly is protein>RNA>DNA the professed informational capability is in the reverse order. As always, and certainly within the context of the evolution of a network rather than an individual, by far the most likely scenario is that of the chicken and the egg, a co-evolutionary process that, with the inclusion of the crucial cell membrane, becomes vastly more probable within the context of the alkaline vent hypothesis.

Robert Shapiro in his sceptical review of abiogenetic speculations, "Origins", strongly favoured the view that perhaps some specialized proteins originally performed the information storage function. He further speculates that the stromatolites which, by analogy with modern bacterial colonies may have, in fact, been formed by other than nucleic acid equipped primitive proteinaceous organisms. Whatever the validity of such a claim, he also draws attention to the fact that the billion or so years of little apparent activity gives a greatly extended time-frame to accommodate selective evolution of the nucleotides. The book "Origins" was published some thirty years ago, however, when the alkaline vents hypothesis was in its infancy, and this much greater time-frame for the emergence of DNA is perhaps no longer required to bring the probability of the rather difficult nucleic acid syntheses within reasonable bounds. In summary, we can safely conclude that the "primordial soup" based models are non-starters. Assembly of functional complex molecules by random collisions in energetically unfavourable conditions really won't wash. An "assembly line" is required. An assembly line with arrangements of jigs and clamps and presses that will enable molecular fragments to be trapped and held in place for welding together. Such arrangements are, in molecular terms, known as catalysts. The catalytic assembly line also needs to be powered. Electrons need to be introduced at just the right places and actively removed elsewhere at lower energy levels. While both the generation and operation of the assembly lines are implemented by stochastic ratchets they are also in some ways "just right". Which should not surprise us at all, because wherever we look in nature, we observe a plenitude of features that are similarly "just right"

# 10

# THE PARABLE OF THE MAGICAL MIGRATING MAT

## An extensible anecdote based upon a true story.

*Once upon a time, in a big palace, there was a large room with many doors that accessed other parts of the building and also had a large window in one wall. In this room there was a large mat which sat in the middle of a carpet. The situation was such that there was frequent random foot traffic across this mat. A strange phenomenon was consistently observed inasmuch as the mat, as if a thing possessed, would, over time, unerringly make its way towards the window. This would recur when the mat was returned to its original position. This strange behaviour was much discussed with various visitors to the palace.*

*One of these, a theologian, pronounced that this miracle was ordained by a deity for its greater glorification.*

*Another, a rather naive biologist, was convinced that this was clearly a result of phototropism associated with some kind of ciliary propulsion.*

*A philosopher of considerable repute declared that this consistent creeping was a clear indication of purposeful behaviour.*

*A young electronics enthusiast suggested that the mat perhaps had perhaps been cleverly designed with concealed photocells and electro-mechanical transducers maintaining the charge on a battery that powered an omni directional traction system.*

*An engineer who then came to visit advanced a far more prosaic interpretation: Simply that the interaction of the pile of the carpet and the textured underside of the rug was such that stochastic inputs provided by transiting feet would cause the ratcheting forward of the rug. By pure happenstance, towards the window.*

*So, in our little fable: The theist simply drew upon unevidenced superstitious mythology, The naive biologist (in no way representative of biologists in general) had jumped to a false conclusion. The philosopher could be said to have had some justification for his stance, in view of the difficulties inherent in pinning down the concept of "purpose" with all its anthropocentric overtones. The electronics novice invoked premeditated design. But the straightforward empirical explanation of the engineer, based on the ratchet, a well established feature of the "big picture" provided to us by the sciences, held the day.*

*The very high level of directionality observed in most natural processes, ranging from the formation of the chemical elements in stars right through to the evolution of technology within the collective imagination of our species lends itself to a similar interpretation, that of inevitable gross outcomes as a result of stochastic inputs acting upon the underlying structure of our universe.*

*In the case of biological evolution, among the more important random components are mutation, the vagaries of gene transfer in the bacterial realm and of the sexual shuffling that has been adopted by many eukaryotes. The ratcheting mechanism here being represented by the process of natural selection.*

*Again, in the phase of stellar nucleosynthesis, although no equivalent of natural selection is easily identifiable, gross outcomes are quite predictable for a system of given mass although a very high stochastic component inevitably pertains in the evolution of more intricate atomic species. Only the most ardent determinist would claim that the evolutionary trajectory of any particular primal hydrogen atom is, even in principle, predictable.*

*Indeed, wherever we look, at least in this stelliferous era in which we exist, we find the machinery of nature that gives rise to increased intricacy is largely probabilistic. Also that the prime motivation for the process can be identified as gravitation.*

*We will explore the evidence for the ratchet in Chapter 19*

# 11

# NETWORKS AND SWARMS

## The Importance of Cooperation

All things within the part of this universe to which we have access are related to each other. But as distance increases, interactions within this universal network are greatly constrained by time. They are also constrained by the means of transmission of information. For very large scales, such as that of the galaxy, the inverse square law for fall-off of photon density from isotropic radiative sources further limits interaction and the possibilities for feedback within such tenuous networks are thereby greatly reduced.

While certain interactions involving feedback can be identified in stellar nucleosynthesis, it is only with the advent of biology that strongly interacting intricate networks of chemistry and geometry come into their own and that feedback rules the roost. In fact it is implicit in any viable definition of biology. For, using the most popular present definition of biological life, imperfect replication and selection, the pattern we see in all organisms today is that of a positive feedback loop.

Replication of genotype
V
Test phenotype in prevailing environment
V
Replication of genotype.

When an error occurs the feedback cycle is reinforced or impeded according to the outcome of the phenotype test. Using an alternative definition (also not exclusive of technology), that of an entity with receptors and effectors along with signal processing in between, the presence of feedback loops is implicit. All but the very simplest logical processors having such a loop.

In the past there has been a tendency to adopt a parochial view of bio-evolution. A mind-set not extending far beyond that of Darwin's "tangled bank". But no species evolves in isolation or even within a very limited domain in time and space. Indeed, it is sometimes very misleading to think in terms of a species or even of a genome. Because, fundamentally, the evolving entity is the biosphere. An intricate network comprising all organisms and certain aspects of our planet. Nested networks of co-evolving chemical and geometric structures. An overall system within which all the components are in some way, to a greater or lesser extent, interdependent. But having a considerable inherent degree of redundancy such that its integrity is strongly buffered against catastrophe. A vast plexus that, from its abiotic inception, since its emergence from a pre-existing network of simpler chemistries and geometries, has further increased in extent and intricacy.

One way of representing this network is as a tree which plots the divergence of species against time. This, of course, is the traditional approach that has been used since the nineteenth century and still inhabits many text-books and popular accounts of bio-evolution. However, while useful for some purposes, it has been found to be a far from accurate representation. Horizontal gene transfer, particularly at the prokaryote level, and sex (zooming in to the sub-species level) are among the various effects that cause branches to become cross-linked. Inculcation of this tree-based paradigm, of which gene ethology is one derivative, had the unfortunate effect of drawing attention away from the perhaps more important web-like network model for which Lyn Margulis, among others. fought long and hard to eventually achieve acceptance.

Sometimes evolutionary pathways have diverged to such an extent that genomic hybridization is no longer physically possible because the genomes ant phenotypes have changed so much as to no longer be either chemically or geometrically (mechanically) compatible.

Nevertheless, interactions between reproductively isolated and genetically distinct branches of the traditional tree are to be found in all epochs. As the planet-wide bio-geo network proliferates, it autonomously generates new niches and is self sustaining and self-amplifying. Constantly "exploring" the possibilities for synthesis of more intricate chemical and geometrical structures from materials both organic and inorganic. Also for harnessing new sources of energy to drive its dysentropic advance. While sunlight is the most obvious source of energy from our human perspective, both the heat gradient and the wide diversity of molecular species generated by endothermic reactions that are abundantly provided by the Earth's mantle play an important role, as does, for instance, the early oxidation of the seas by solar radiation which enabled much chem-autotrophy.

The dramatic advances in genetics around the middle of the last century had the unfortunate effect of nurturing a paradigm which placed far too much focus on the gene mutation as the source of evolutionary diversity. Often naively portrayed as simply the result of competition between genes. The clearly evident reality of the high level of close interdependence and cooperation observed in the network having been carelessly swept aside. With only a few hard-nosed biologists, of which Margulis was a notable example, keeping their feet on the ground and retaining a considerably more balanced perspective. The genome, when all is said and done, is merely the database, or what in earlier times we would have called the reference library, to which the cell can refer. There is no question as to its indispensability for at least most cells. But even at the level of a single cell, it is only one small component of a complex cooperative system. The model of biological evolution that portrays the process as random mutation of genes producing phenotypes that are filtered (mostly deleted) by natural selection is not wrong.

Like the Bohr model of the atom, it is useful as a first approximation to provide an introduction to the subject. But it is very far from complete. For example:

- The influence on an individual, because of the complex interactions, competitive and cooperative, with other genes is seldom, in the larger multicellular organisms, directly identifiable with specific inheritable traits in the phenotype. Particularly in the case of more complex organisms, the situation is more that of a member of an orchestra playing in symphony rather than a soloist. The genome is most certainly not the conductor, nor did it write the music.

- Genes can be transferred from one organism to another. It is now incontrovertibly established that such transfers are quite commonplace among prokaryotes and that, even between members of the far more divergent multicellular phyla, transmission of plasmids by viral vectors quite often does occur. Giving a rich source of variability, quite distinct from that of genetic mutation, upon which natural selection can play its usual part as grim reaper.

- Those organisms that include sex in their repertoire provide yet another valuable source of variability upon which selection can act. We should bear in mind that the many animal breeding exploits carried out by our species are not really accurately described as genetic. These efforts, instead, are merely manipulations of the phenotype using inherited traits.

- Not all genetic changes are adaptive. In other words they are not driven by selection pressure. Rather, those not disallowed by nature's gatekeeper diversify by random diffusion within a pre-existing niche. This is common among the flowering plants. My garden accommodates a huge range of flowers having strikingly different colours and morphologies. Many of which quite happily share the same pollinators, the same soil-

> based nutrients and all enjoy the "free lunch" provided by solar photons. While more competitive environmental histories and, more recently, horticulture may have played some part in this, genetic drift within a niche accounts for much of biological variety. It is an area where contingency really comes into its own during the initial early phase of completedepascency. The effect being particularly strong where populations, for one reason or another, become isolated. The accommodation of numerous devices that can do the same job within one niche adds redundancy to the biospheric network and thus increases its resilience.

Other sources of variability include epigenetic effects such as methylation and acetylation of nucleic acids during an organism's life cycle, which some current research suggests, can sometimes be passed on to subsequent generations. Other considerations are viral interventions and events of uncertain origin in which the gene, chromosome or whole genome are doubled.

Quite apart from a gross oversimplification of the basis of biological diversification, the twentieth century bout of "gene-fever" generated a seriously distorted view of biology's basic units, the cell and the phenotype. The primary reason that other important components of biological machinery became subsumed by the gene and the genome would seem to be one of practicality. The nucleic acids are, in fact, relatively simple and rather boring molecules. Rather long, of course, and without the resorting to the use of pre-existing cellular tools, fairly hard to synthesize. By cheating a little, however, and using these very handy enzyme tools that are provided by the cell, nucleic acid research became a rather attractive option. Spurred on by the feedback loop provided by its lending itself easily to automation. The big money to be made in the manufacture and sale of ever more sophisticated and easy to use instrumentation resulting in intensive marketing of these products. Scientists, of course, are not immune to marketing pressures, especially when holding the promise of making this kind of research a breeze. Fanning the flames of gene fever even further were the numerous practical and very useful applications springing from such work. They include forensic, medical and industrial uses as well as helping to extend our knowledge of nature by, for instance, estimating evolutionary distances.

Among this great distraction, the genetic hyperbole with which even the layman is constantly bombarded, the real unit of biology lies largely forgotten. It is the cell, the co-operative network of which the genome is but one very tiny component. Even the bacterial cell is incredibly complex. Far more so than DNA. The cell of the eukaryote is of mind-boggling intricacy. It is a network whose fine structure, let alone detailed functionality, is still far from entirely understood. All our observations show that a cell can only come from a cell and is simply a result of division. The two half size cells thus usually produced, having, in almost all cases, exactly the same components as the parent. Each being an independent and self-sufficient machine that can "hit the ground running". Each "daughter" directly inheriting all compartmental structures, ribosomes, mitochondria and other organelles together with the full enzymatic tool kit from the parent cell. The huge amount of information this represents is passed down the line ***without*** the direct intervention of DNA. It is only for subsequent growth of the individual cells that the protein co-operative needs to refer to the genomic library for information as to the sequences which are required to enable the ribosome to assemble amino acids into suitable proteins. Proteins that the intracellular machinery, with its vast array of enzymes (which include the directly inherited family of RNAs) can translocate into structures or restock the enyzmic toolbox as appropriate.

In May 2010 the public were bombarded with hype to the effect that *"Craig Venter creates synthetic life".* We have come to expect such arrant nonsense from the media, of course. But the following comment attributed to Venter is most unbecoming of a scientist:

> *"This is the first synthetic cell that's been made, and we call it synthetic because the cell is totally derived from a synthetic chromosome, made with four bottles of chemicals on a chemical synthesizer, starting with information in a computer,"*

In fact, far from being "derived from a synthetic chromosome", a perfectly standard naturally derived common or garden bacterial cell, with all its inherent complex machinery and structure intact, had merely been modified by swapping its original protein specification database for another. While this was no mean feat in itself, and an intervention possibly having practical potential, this was a trivial endeavour compared with building, rather than simply modifying a cell. More recently a team at the Scripps Research Institute have introduced a pair of artificial nucleotide bases to *E. coli* DNA, binding by hydrophobic interactions rather than hydrogen bonds. While the cells thus modified are viable, the claims that this approach will lead to the production of novel proteins would seem to have little substance since the cell has no machinery to process these additional bases.

Nevertheless, the very fact that a nucleus can be temporarily removed in this way (as also occurs in the process of cloning) to allow for the exchange is in itself evidence that a cell can, at least briefly, stay alive without its genome. A fact for which we can find very good support not far from home, right in our own bodies. The red blood cells of mammals have been trimmed right down to minimum size and maximum flexibility so as to allow access to fine capillaries. After the protein assembly required for their initial growth in bone marrow is complete, the whole nucleus is discarded. Only the barest essentials of cellular machinery remain. Even the ubiquitous mitochondria have gone, their respiratory function replaced by direct glycolysis akin to that occurring aerobically in muscle. With absolutely no DNA, the red blood cells work just fine. The co-operative efforts of the remaining support systems providing smooth operation over their life-span of a hundred days or so.

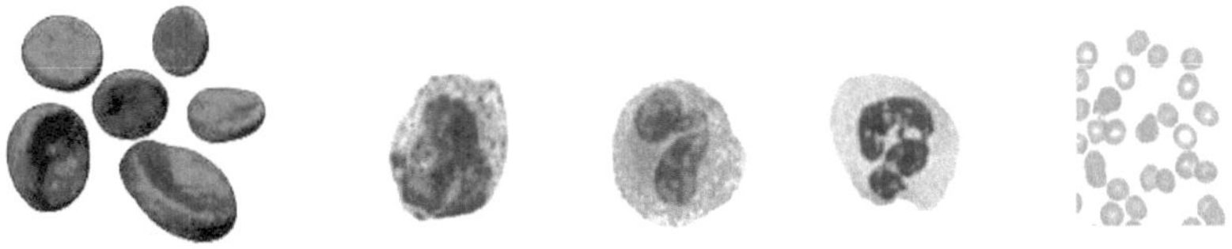

Erythrocytes Leukocytes Platelets

Yet another example is provided by the platelet cells, which are involved in blood clotting. These have most of the usually present cell machinery, now including mitichondria, cytoskeleton, surface glycoproteins and microtubular system. Again, here we have adult cells which, having relinquished the nucleus after their initial bone marrow growth stage, are entirely viable, highly organised and fully functional, with the full complement of enzymes running the show. They are, however, totally devoid of nuclear genetic material and ribosomes. Why? Simply because they are single-use cells. They will not have cause to grow further or divide, so have no requirement for components whose only function is the provision of recipes for regrowth and replacement.

The third main category of blood cells, leukocytes, the military divisions of the community of cells, present an interesting problem.

For, although they, too, are disposable and short-lived, typically surviving around five days, they come fully equipped with a nucleus. Our engineer might guess that this is required for the on-going manufacture of enzymes for lysing invaders. But, at least with the eosinophils and neutrophils, this seems not to be the case. The set of enzymes they start off with not being replenished. It appears that the DNA is in this case is actually used for a very different purpose. More aptly compared to a fishing net than a recipe bank. For it is extruded from the cell to ensnare the enemy. Again, other kinds of leukocytes are destined to become stationed in tissues to perform their defensive duties. These use their nuclear DNA for conventional roles in growth and changes in shape. Excellent examples of nature's cunning.

So we see that, contrary to popular opinion, in no way is the genome the master or the "brain" of the cell. It is but one of myriad equally important elements in a highly interactive co-operative network.

Humans seem to be irresistibly drawn to hierarchical models. Perhaps because we are brought up to view our social structures, and those of some other animals, in this way. We like to think of our brains as running our bodies, the queen bee as running the hive, DNA as running the cell and, inevitably, a deity running biology. Whereas all of these are more properly regarded as network functions. Hubs, rather than hierarchies are the rule in natural processes.

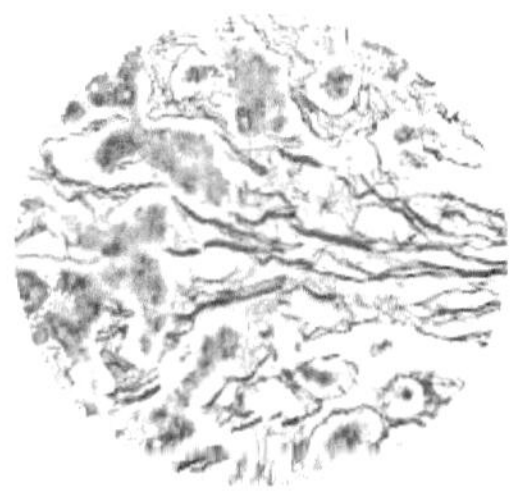
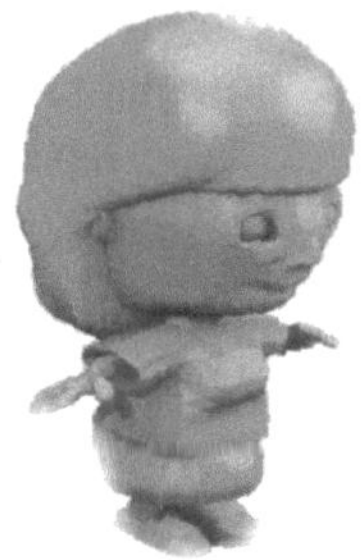

Myoglobin 2MB Liver tissue Girl 2MB

We see above that even the very crude representations of the protein myoglobin, which is a model generated by a 3D printer, each require 2MB. The models shown are printed in monochrome polylactate and are homogeneous. At an atomic resolution we can expect the data requirement to be many orders of magnitude higher. The centre image is a micro-photograph of a section of liver to illustrate the extremely high levels of morphological and functionality intricacy at all levels. In the liver alone it is estimated that each cell contains somewhere around 10,000 unique proteins, a goodly proportion of which are enzymes, and a total population that can run into millions. With a couple of billion cells comprising the entire organ. In the human as a whole there are over 200 well-identified cell types, with an overall population of some 30 trillion cells.

Now, the total data encoded in the base pair sequence of the human diploid genome works out around 3 Gb (~0.4 GB). This is not a lot of information. Only enough to specify the print-out of 500 different plastic protein molecules, or dolls. Not even enough to fill more than a single CD. Enough only to fill ten two meter library shelves. A level of intricacy that equates to handful of Beethoven symphonies.

Clearly there has to be more to generating an organism whose brain alone is widely touted as the "most complex entity in the universe". The fact that this claim is wildly wrong is here of no great consequence because, even allowing for the hype, the human brain ***is*** pretty damn complex. With vastly more intricacy than can be squeezed on to a couple of CDs, even if we very reasonably credit nature with evolving optimal data compression strategies. So how are we to escape from this enigma? The answer, as in all things biological, is to be found in the distributed properties of the network rather than any particular component. It can help our intuition if we use a piano keyboard as a metaphor for the genome. The piano has 88 keys, which, in our present context, we can treat as 88 bits of information. But, of course, even using just one finger, we can play thousands of different tunes. What's more, we can also play chords, which adds a whole new dimension to the possibilities. The available combinations and permutations are, for practical purposes, endless. Viewing the genome in this more informed light, then, we see that assembling an animal comparable to us, or an oak tree or a blue whale should be an absolute breeze!

So where's the catch? The catch is that our piano requires a player. Exactly as with the genome, it is no damn use our baby grand or honkytonk just sitting there gathering dust. For its function it needs to interact with the network, and the more complex is the output required, the more complex are the input requirements. For our DNA piano, the input is supplied by a feature of the biosphere that we can describe as a musician. It is true, of course, that a piano can output by itself if it is configured as a pianola, or "self-playing piano". Items which enjoyed a wave of popularity in the early twentieth century. But in that case the roll of punched paper which provided the note sequences was merely a surrogate musician with its structural and informational roots extending deep into the biosphere. Where, then, is there to be found the "player" of the genomic piano? Where else but in the network that is the cell itself. Within which has been directly inherited, by mitotic division, all the intricate machinery required for knocking out the right kinds of tunes from its keyboard, its protein recipe database. For the playing of this DNA piano we have, among many other enzymic bits of kit, transcription factors such as "enhancers" and "silencers" that pick out the right chords as determined by the stream of signals constantly arriving from the intracellular, somatic and extra-corporeal networks.

Accidents happen. Pianos can sustain damage, strings can go out of tune such that when the musician plays the right keys, the wrong notes are output. Very occasionally, in the case of the genomic piano, similar accidents are bound to occur, and, even more occasionally, the tune produced becomes a hit, an improvement on the original, fits the phenotype better to the current state of the biosphere, and is actively conserved by selection. The human pianist is not just a bunch of fingers and thumbs. She is an incredibly intricate community of cells, a coherent swarm. Most of the individuals of which, neurons , liver cells. blood cells, lung cells, heart cells and so forth all have to be functioning properly to enable her proper musical output. The cell, too, is a swarm. A swarm of specialised molecules and modules, tools and machines, if you will. All of them performing many and varied tasks. Playing the genomic piano is but one of these. The fingers and thumbs for this task are best represented by some of the variants of RNA. The family of microRNAs. Which, of course, are either directly inherited or synthesised as required by the abundance of extremely clever proteinaceous enzymes. As are the fats, sugars, steroids, phospholipids and all manner of other molecules that are required for local repair, maintenance and growth. As well as others, specific to location and function, for exporting to the far wider somatic web. The intracellular collective even builds new pianos to bequeath to the next generation. For the genome does not duplicate itself – that tricky feat is engineered by dedicated intracellular machinery. Machinery that has been either inherited directly or manufactured *in situ.*

We have looked at some of the shortest lived cells, the erythocytes and platelets, which, having portability and disposability among their requirements, discard DNA on reaching adulthood. At the other extreme we have neurons, most of whom are wired in for life and do not replicate. Because of their longevity, however, the protein recipes are still required for routine maintenance. Something very interesting and rather unexpected, though, has been discovered in the course of single neuron sequencing carried out by Fred Gage and his team at the Salk institute. The genomes of individual neurons turned out to be far from identical. One explanation for this could simply be an accumulation of failures of the normal DNA repair mechanisms over the longer time-span. This hypothesis suffers from two weaknesses. Firstly, extensive random disrepair we could expect to be reflected in pathologies. Indeed, brain cells do seem to eventually "wear out". But senescence is common to all body parts and more likely attributable to mitochondrial dysfunction. Secondly, the variations have the hallmark of constructive transpositions rather than random disruptions. Which raises the possibility of pro active modification of the genome by the intracellular machinery, perhaps an exaption to aid information storage. Certainly evidence of methylation of DNA to this effect has been advanced by Miller and Sweatt off the University of Alabama. They find that:

> *"DNA methylation is dynamically regulated in the adult nervous system and that this cellular mechanism is a crucial step in memory formation."*

Turning to the special case of sexual reproduction, the simple division of mitosis which, although more involved, is not too far removed from the binary fission of prokaryotes, the more complicated procedure of meiosis is employed by the cellular machinery. Enzymes separate the pairs of homologous chromosomes derived from the two parents and redistribute gene sequences between the two. Two rounds of division occur, the second involving a separation of the homologous chromosomes so that only one remains in each resulting gametic cell. Those destined to become sperm are stripped down to the bare essentials that enable it to have a chance of delivering its single chromosome set to an egg. The egg, on the other hand, also having a single chromosome set, is equipped with the full complement of cellular machinery to read the revised database produced by the merger. The cell is in this way presented with a different combination of recipes that provide variations on a theme. Since the individual protein recipes are all well tried and tested, however, far less risk is involved in the tweaking of cell properties than in the case of mutation.

Other ingenious catalytic moieties, having received instructions passed down the line from the various navigational sensors that detect the position and status of the cell within the somatic web, carefully apply recipes read from the genome so that the raw proteins that are generated by the ribosome meet the specific requirements for that particular tissue. In addition to the strategies mentioned previously, another known way in which this is done is to activate or suppress appropriate gene sequences by methylation of parts of the genome itself, or acetylation of the histones which provide support. In this way the correct raw materials for the role which that kind of cell plays are custom-made. But the work of the intracellular network does not stop there. For it's no use to just have a ribosome dumping a great pile of raw proteins willy-nilly on the factory floor. Care needs to be taken to ensure that they become neatly folded into the shapes that are vital to their function. They also need to be translocated for incorporation into structures or, if they are enzymes, installed properly at their appropriate work-stations. The cell is a hive of information-rich molecular industry employing a skilled and very organised work-force. A work-force which has inherited the entire wealth of information that is transmitted directly from cell to daughter cells. An amount of information greatly exceeding that stored in the tiny protein library of DNA! When an individual cell line reaches its use-by date it dies out. But all is not lost, for new cell-lines are set up, Always complete with the fully functional basic network of the appropriate kind of stem cell.

Within most eukaryoyic cells, too, are a variety of organelles , each of these having, within their confines, their own special set of highly networked cooperatives, their own individual swarms. And in case of the mitochondria and chloroplasts, their own local branch library of protein recipes. None of these entities totally isolated, of course, but interacting with all other cell components via their membrane gateways, just as the cell itself interacts with the bodily network. Exosomes, tiny packets containing RNA and proteins which are exported from the cell are but one example of the many agents which are currently under scrutiny as mediators of such communication. The bodily swarm, in turn, interacts with outer networks, such as the biosphere and, most notably in the special case of the language-enhanced humans, the technosphere. It appears that it is from such highly networked interactions, powered ultimately by gravity, that all biological intricacy arises.

So we see that the perception which is at present still abroad that the genome, despite its simplicity and relatively small size, alone determines the morphology and functionality of the phenotype is patently invalid. The "central dogma" that information flows only from nucleic acids to proteins is contradicted by the experimentally observed methylation and acetylation of the genome. This is certainly an information flow from protein to genome. Indeed, this process also involves appreciable information flow from the extra-somatic environment right through to the genome. That such epigenetic modifications are not, as a rule, carried through to further generations in no way negates the validity of this "reverse flow". In actuality, despite the routine "clean-up" of the genome while it is being duplicated, it is becoming increasingly evident that in certain instances some of the epigenetic modifications ***can*** be carried through to further generations. The biggest mystery being by what mechanism the modified genome could become incorporated into the gametes, which, in sexual creatures, seem to be tucked away safely at an early stage of development. However, we should not overlook the rather sacrilegious possibility that the revised information might instead be inherited directly from the parent oocyte.

Whether or not this indubitably contentious suggestion should have any validity, the far more significant point is that the most essential channel for inheritance, for information transmission, is direct transfer of machinery from parent cell to newborn progeny. Here we have what amounts to a massive flow of information, effectively from protein to protein (and, for that matter, phospholipid to phospholipid, etcetera). This, in itself, a clear falsification of the central dogma hypothesis.

That this really rather self-evident primary mechanism of inheritance has been generally neglected for so long is disturbing. The few dissenters from the gene-centric obsession, such as Franklin M Harold, now at last find their more realistic perspectives robustly vindicated by experiment. The recently burgeoning interest in the micro-RNAs, has led to work by Craig Mello and others whereby the introduction of one of the regulatory micro RNAs to the cell not only changed the status of the immediate phenotype but this change proved fully heritable though many generations. It also lends some weight to Robert Shapiro's suggestion that the database function might have originally been effected by specialised proteins. Perhaps a rather clumsy arrangement which was eventually abandoned in favour of a more compact and efficient nucleic acid based recipe library derived originally from RNAs which had taken over particular enzymatic duties.. Possibly derived from some of the related monomers commonly used in metabolic pathways. A surprising complete reversal of the DNA>RNA>protein dogma! A possibility which, if correct, might possibly cast the prion in the role of a relic of that primordial arrangement.

That, however, is speculation. The present bilateral flow of information between genome and protein is firmly based in experiment. As is the very obvious but often ignored direct transfusion from parent cell to progeny by division. Also under-appreciated is the bilateral flow of all many kinds of information between the swarms at all levels of the biospheric network. The emergence of language added a whole new dimension to the swarming behaviour of the human animal, but we will leave discussion of this until later chapters.

The development of a zygote into a soma, a fertilized egg into an organism, compares closely to the overall evolution of our universe in microcosm. An example of a highly directional evolutionary process. In this case, although it seems that competition and selection both evidently take place at early stages of gestation, perhaps due to nuclear and mitochondrial incompatibilities. A chicken, or a human baby is primarily a co-operative venture. A process guided by the cellular machinery provided by the mother in response to signals of intracellular and extracellular origin. In conjunction with access to the protein recipes stored passively in the well shuffled DNA.

Recognition of the essential similarity between the networks that correspond to phases of the overall evolutionary process is becoming evident from all quarters. For instance, Dmitri Krioukov, of the University of California, in a recent open access paper observes:

> *"Structural and dynamical similarities of different real networks suggest that some universal laws might accurately describe the dynamics of these networks, albeit the nature and common origin of such laws remain elusive.*
> *Here we show that the causal network representing the large-scale structure of space-time in our accelerating universe is a power-law graph with strong clustering, similar to many complex networks such as the Internet, social , or biological networks.*
> *We prove that this structural similarity is a consequence of the asymptotic equivalence between the large-scale growth dynamics of complex networks and causal networks. This equivalence suggests that unexpectedly similar laws govern the dynamics of complex networks and space-time in the universe, with implications to network science and cosmology."*

The realisation that the evolution of our universe is best described in terms of a network of ever-increasing intricacy, and that its phases, sometimes overlapping, are contiguous is at long last coming of age.
The advent of sex can be broadly regarded as a response to the increasing size, intricacy and more rapid rate of change of the environmental network. A self perpetuating and exponentially amplifying positive feed-back loop. Thrust upon the biosphere by the replicative requirement for greater flexibility demanded by the eukaryote, particularly the multicellular implementations. Not so much an option as a necessity.

There are several kinds of argument raised by those who would deny the directionality of evolutionary processes. The ancestral loss of legs in the snake or the dolphin ancestors are among those most often quoted. Certainly such changes occur. Where there is a niche to be filled, the pressures of natural selection will find it. The changes are not, however, devolutionary. The advent of the snake, the dolphin and the visually challenged mole all increase the overall intricacy of the network. A more interesting example involves the viruses. A class of structures that are not widely considered to be "living". But, without a shadow of doubt, are part of the evolutionary "life" process.

The jury is still out on the question of at what stage the viruses joined the biospheric network. Indeed, it is not inconceivable that some may have different ancestries from cellular organisms and acquired (or donated) the nucleic acid commonality by interactions in the process of co-evolution with the ancestors of the archea and the bacteria. Certainly the viruses greatly outnumber their hosts today. The main competing and more favoured hypothesis is that some (or all) arose by a "retrograde" evolution of cellular organisms. By dispensing with cell membranes and associated machinery to produce, in the most extreme implementation, the viroid, a simple bare-bones data-set for hijacking cells, and keys to unlock their barriers. This is the implication irrationally adopted by those in directional denial. We can repudiate the "no direction" argument here in exactly the same way as we have with the snake and the mole: The loss of intricacy of the organism in respect to its ancestors is more than compensated for by the addition of a new niche. An overall increase in intricacy for the network.

We can look at this from another perspective, too: The virus, or any other parasite, for that matter, is entirely dependent on its host. Without a host species it is impotent. So, in determining intricacy, it is not sufficient to look at the parasite alone. To attribute any useful meaning we have to consider the parasite-host system in its entirety. Which is obviously far more intricate than the host in itself. This reasoning applies even to the very striking example of "retrograde evolution" demonstrated by a team led by Sol Speigelman. They successfully persuaded a virus (whose natural host is an *E.coli* bacterium) to instead replicate in a solution containing a replicase enzyme from the bacterium together with suitable nutrients. The virus flourished, in fact faring much better than when in its natural hosts. Because whenever a mutation occurred such that a gene sequence concerned specifically with its usual interactions with its host was lost, the virus could still carry on its merry way in the more congenial niche very kindly provided by Sol and his pals. Evolving within this cushy artificial environment, many more redundant nucleotides were shed, the essential virus structure becoming ever less intricate. The total number of nucleotides being eventually reduced from about 4500 right down to 220. These retrograde replicative entities, stripped right down to their bare essentials came to be widely known as "Spiegelman's monsters". The ultimate "couch potatoes"

So surely, it is argued, evolution is here producing a decrease in intricacy. Those who arrive at such a conclusion have simply not thought it through carefully enough. They have fallen into a trap akin to that to which many succumb with regard to entropy. Certainly these 220 unit monsters are far less intricate than their "all bells and whistles" predecessors. We must admit that the ***local*** intricacy of this system has decreased. But what has happened to the ***global*** intricacy? The monster is no longer capable of inhabiting its predecessor's bacterial host. It is now utterly dependent on a new host. And what is this new host? It comprises an intricate network of such things as the glassware in which it resides, the apparatus that controls its temperature and nutrient concentrations, the laboratory in which the work is conducted, the behaviours and imaginations of Spiegelman's research team, as well as the vast infrastructure of the extended imaginations of humankind! Now, very little in nature is simple. And even an *E-coli* bacterium, it must be admitted, is an enormously complicated bit of machinery. But certainly nowhere near as intricate that of the vast network of mind and matter that has enabled the existence of Sol's wee monster.

Prompted by his own experimental results, Joe Thornton of Chicago

University provides similar examples :

> *"...complexity increased because protein functions were lost, not gained," Thornton said. "Just as in society, complexity increases when individuals and institutions forget how to be generalists and come to depend on specialists with increasingly narrow capacities"*

There are, of course, extinctions, once viable niches which for one reason or another cease to exist. There are catastrophes. Such as the great Chicxulub asteroid impact, 65 million years back. Believed to be the cause of rapid extinction of a large proportion of species, particularly those immediately dependent upon photosynthesis. The eruption of Toba, about 75,000 years ago, while not of the same scale, is believed to have wiped out many species, including a substantial proportion of humans. But natural dynamic networks are remarkably resilient. Even that which is the community of cells comprising our bodies can often recover from horrendous injuries and, in these times, extreme surgery. Our Internet having evolved along somewhat similar lines, with multiple cross-linkages, redundancies and distributed functions, it is both resistant to attack and is self-healing. We see the phenomenon again and again in social structures. Japan and Germany, severely ravaged by war, recovered and rebuilt into new niches and now bloom and prosper.

Despite immensely damaging floods, earthquakes, fire-storms, eruptions, economic meltdowns, these networks have an amazing capacity to pick themselves up, brush themselves down and get right back in the game. Similarly the mass extinction events such as that called "The Great Dying" at the Permian/Triassic boundary maybe set the biospheric network back on its heels for a while (as have, no doubt, the various ice-ages), but in the longer term they supported an acceleration of its evolution by virtue of the great variety of new niches that subsequently eventuated. There is a lesson to be learned here. Contingencies may inhibit network growth for a while and change the paths by which it proceeds. But because the network is self-healing and self-correcting, similar gross outcomes are attained.

Another argument used in an attempt to discredit directionality points to the increasingly greater branching that occurs as time progresses. Sometimes expressed as "it is not an evolutionary tree, but a bush". This argument is self-defeating because that ever higher degree of branching exactly reflects the ever increasing intricacy of the network. Observed even within the limited context of the traditional tree diagram, where the very important lateral interactions are left out of the picture.

So far we have been looking at the biospheric network, to which we will return in a later chapter. But there is another network with very similar characteristics: That representing the evolution of technology. Which takes place within the collective imagination of human minds in the form of ideas and is manifested physically by artefacts. Artefacts by the billions. Generated by the recombination and synthesis of ideas which ultimately draw upon the environment for their raw materials. The data input from the world by our senses. Shared among minds and harnessing the power of the swarm by means of language. The unique large-scale import, export and external storage of imagination. Itself an inevitable consequence of evolution of the human animal which has been brewing in an ancestral network for at least two million years.

# 12

# GRAVITY

Albert Einstein once remarked:

"Gravitation cannot be held responsible for people falling in love."

But in this he was very wrong! Gravity appears to be the motivator of all intricacy. Including, of course, our own biochemical mechanisms.

Seldom acknowledged is the undeniable fact that the ultimate source of all our energy, at least in this stelliferous era, is gravity.

All engines that on Earth do dwell, including biological ones, are powered, directly or indirectly by this phenomenon.

The water-wheel and the clock of previous generations, in which a bob-weight rather than a spring was used to store energy having a very direct involvement, and sometimes thought of as "gravity driven".

This view is rather deceptive, though. Similarly, it is true that a waterwheel is driven by water flowing down a gravitational gradient. And that the rain which replenishes the river falls by gravitational attraction. But, again, we can *trace the energy back to that which evaporated water to rise against the Earth's gravitational field and form clouds which coalesced to form the rain which fell from the blue.*

*So now we get back those carriers of energy, incoming photons, which caused the evaporation, as the next suspects. But we must not shoot the photons, for they are mere messengers. Carrying packets of energy from the sun. And where did the sun get this booty? From gravitational collapse!*

*Whatever power sources we choose to consider, we always find a trail leading back to the same place. In the example of the clock the human animal has merely stored, in the raised bob or the spring, some of the energy he or she has acquired by eating plants (or other animals that have photosynthetic organisms at the bottom of the food chain).*

*Again, energy from the sun. Photons in the visible range are the carriers, this time. But here the trails converge and gravity is found at the root.*

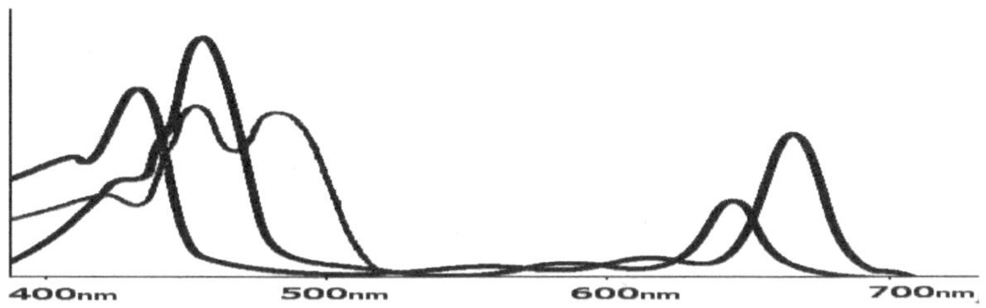


Photosynthetic absorption maxima

*And this is always the case. Wind power and tidal power, too. The fuels we commonly use to run our automobiles, household and industrial appliances, oil, coal and so forth, these all go back through the photosynthesis path.*

At first glance, hydrothermal energy and that used by organisms having sub-oceanic habitats involving "black smoker" volcanoes or fissures would seem to be exceptions.

But, although the trail does not involve our sun, it still leads back to gravitation. . The collapse involved in formation of the planet contributing to the high temperature of its core. Also the contribution provided by tidal effects.

Similarly with the additional heating therein (and also the power generated in our nuclear fission power stations). The radioactive fuel for which, uranium, finds its ancestry in the gravitational collapses that generate supernovae.

Tidal power

Direct gravitation
Interaction of masses
High metallicity
Supernova
Gravitational collapse

Geothermal power

Fissile materials (mostly)
Supernovae
Large star
Gravitational collapse

Nuclear Power

Fissile materials
Supernova
Large star
Gravitational collapse

Ox Wheel
Glucose from plants
Photosynthesis
Our sun
Nuclear fusion
Gravitational collapse

Power station
Fossil fuels
Plant residues
Photosynthesis
Our sun
Nuclear fusion
Gravitational collapse

Hydroelectric dam
Solar heating of water
Our sun
Nuclear fusion
Gravitational collapse

Wind power
Solar heating of atmosphere
Our sun
Nuclear fusion
Gravitational collapse

Wind power
Solar heating of atmosphere
Our sun
Nuclear fusion
Gravitational collapse

Solar photoelectric array
Our sun
Nuclear fusion
Gravitational collapse

# 13

# INFORMATION

It is often overlooked that information is always manifested by an arrangement of physical states. In practice, it corresponds exactly to the increase in intricacy which is the subject of this book. Information is not, as those who choose to inhabit a platonic dream-world might have us believe, some kind of nebulous platonic entity arising from mathematical theory. Instead, it is securely rooted in the the practicalities of the real world of the hard-nosed empiricist.

On your hard drive it is an ordered population of regions having either of two states of magnetization. On a vinyl phonograph record it is embedded in the geometry of the spiral groove. In the DNA of each of your cells it corresponds to the ordered arrangement of four different nitrogenous bases along a strand of phosphate-linked sugar molecules. These are a few of the myriad ways in which information is stored.

To do anything with stored information, to access information from your hard drive, to reproduce the information in the groove as musical sounds, to generate a new biological cell, we always have to expend energy. Mechanical energy that turns the motors in the first two instances. Chemical energy in the case of the cell. Despite using the exquisitely efficient biochemical mechanisms that have evolved for the complex functions of the cell, some of the energy is still wasted as heat. It turns out that this is inevitable. Indeed, we learned of this way back in the middle of the nineteenth century when folk first fooled around with steam engines and eventually managed to figure out the ways of nature that are today known as thermodynamics. Even making their outcomes easily calculable.

Conveyance of information can be effected chemically. The hormone testosterone and the neurotransmitter acetylcholine, both very close-to-home features of our nervous system, are good examples. It can also be effected mechanically. As with the air pressure variation we call sound and by hieroglyphics and so forth. Or electrically, by varying the propagation of electrons along a conductor, as in the telegraph and the telephone. Even more widespread in nature is the use of photons, by virtue of their intrinsic frequency coding, for the conveyance of intricacy from the sun to our planet. Also, of course, often used by ourselves to convey information by means of streams of photons modulated or filtered in a variety of ways. As in the case of radio communications and traffic lights.

So the great evolutionary process of which we are part, a network of burgeoning intricacy, must necessarily be identified with a body of information. To build that structure, an input of energy was required. And, as we have seen, all energy appears to be derived from gravity. The inescapable conclusion being that within the time-scale we are considering, from the birth of the first stars to the present, information is derived almost entirely from gravity. That information and energy are interconversible. And, sure enough, whenever information is degraded, for instance when an organism dies, heat is released from the evolutionary network. In fact, in accordance with the perspective provided by the second law of thermodynamics, change in the structure of the network of any kind is accompanied by a global increase in entropy. Thus we could say that nature extracts a tax on all transactions.

1. In this context it should not be forgotten that magnetic fields are a property of matter and limited to matter having particular kinds of intricacy. Matter having peculiar intrinsic atomic configurations and geometries. Or bulk properties, such as that of the conductor, which allows a flow of electrons to occur. Since we find that intricacy, which can also be described as information, has seemingly evolved from a pre-stelliferous gravitational continuum, seeded by a few atomic species of

> minimal intricacy, we can deduce that magnetism, too, is an emergent property of the evolutionary process. And another aspect of information. The separation of charge beyond that found in hydrogen is yet another phenomenon that can only be attributed to the action of gravity. It, too, is inextricably associated with the geometry of atoms.

The only significant carrier of information not ***directly*** associated with matter (assemblages of atoms) is the photon. We, of course, use modulated streams of photon for conveying information in all our radio-communications. But an individual photon carries only one kind of information and that is colour (more accurately, frequency or its inverse, wavelength). The colour is also a measure of the energy associated with it. Blue more energetic than red. X-rays far more energetic than microwaves.

Furthermore its colour can be changed. A yellow photon leaving the Sun is subjected to a gravitational field. It loses energy. Because, according to our present knowledge, the speed of light in free space is constant, it can't slow down. Instead the information content is changed and it becomes "redder". If it is heading this way, on encountering Earth's gravity its energy will start to increase and the colour will shift back towards the yellow. We again find evidence of the equivalence of energy, information and intricacy. Almost all of the photons, those "mass-less" carriers of information, that we observe today emanate from the cosmic family of material bodies and hence, ultimately, again, from gravity. The only exception being that rather small proportion in the microwave spectrum that are today attributed to the "big bang".

So we find that an alternative title for this book could be "The Information Generator". And that all the objects we see around us can be viewed as information networks. In fact that anything that impinges on our senses, such as the summer breeze that wafts past me as I write, or a brick, or a table, is an information network. Because all of these are intricate structures produced by the evolutionary processes driven by gravity. Certain aspects of, say, the brick, are transmitted to my retinas, filtered, interpreted, and encoded within my neural architecture. For this transfer of information to occur in accordance with the second law of thermodynamics, energy must be expended and thus global entropy increased. The brick is at present en-swathed with photons of the visual spectrum arriving from the sun. Those "reflected" along certain paths then encountering pigments in my retinae. The interaction of these photons with the retinal pigments raise certain atomic energy levels in the pigments together with the effective re-emission of the "used" photons, red shifted to a much lower energy level. Initiating a complex chain of events that culminate in my mental image. And at each step in the chain some energy is wasted and entropy increased. This is why brains, while using very little mechanical energy, run hot and need very effective cooling systems to dump the waste "red-shifted" heat.

That is, of course, by no means all the information that can be transferred to my mind from the brick. I can break the brick apart to get an idea of its internal structure. And if I am seriously interested in extracting detailed information, I have, these days, many ingenious means at my disposal to so do. Various kinds of microscope, x-ray diffraction studies, and atomic absorption spectrophotometry to mention but a few. Intrinsically much more costly, of course, in energy and thus in its close equivalent, dollars. As well as the zephyr, the brick and the table, I spy, with my little eye, a book, a rose and a pussy-cat.

The book is an example of the kind of object which has had extrinsic information impressed upon the intrinsic intricacy of its basic structure. A class of object which has blossomed prolifically since its likely early implementation on a section of cave wall. It includes the papyrus, the abacus, the compact disk, the flash card.

The lovely rose is a whole new ball game. It not only has the information that is identical to the intricacy of its structure but also has sensors for data input, effectors for data output and processing mechanisms between. Something that we certainly do not recognize in the brick or the book. The rose also uses information impressed upon it by natural selection. Such as its sweet perfume and its beautiful bloom. Advertisements for cooperation with pollinators on the immediate network. The rose can be used for transmitting other kinds of information, too. I can cut it and take it to my wife to let her know I love her.

As for puss, we now have an extremely intricate furry entity equipped with myriad sensors and an information processing system that is able to support both a substantial amount of imagination and enormously sophisticated motor control. Very effective for activities such as climbing trees, catching mice and rabbits. As well as for manipulating their pet humans.

There are also some extremely interesting issues concerning the fundamentals of physics that arise from examining the observed interconversion of gravity into intricacy, or its exact equivalent, information. The reader may have noticed that in the interests of simplicity, I have previously set aside the other physical forces admitted of physics today. But perhaps it is time to check these out:

The weak nuclear force, of which magnetism, according to the most common electro-weak interpretation, can be treated as an intricacy-related manifestation. We can safely make that assumption because magnetism, in each of the three "flavours" in which it is manifested, is observed to be associated only with particular atomic configurations. Certain levels of geometric intricacy.

Lastly, we consider the strong force, that which binds together protons and neutrons (as well as other hadrons) in the nucleus of atoms. Since there is no strong force present in the hydrogen atom with its single (unbound) proton, and the hydrogen atom is the primary building block of all the atoms produced in the generation of intricacy it would seem to almost inevitably follow that the energy/information found in atomic nuclei having many protons and neutrons is also derived from gravity. Unless, perhaps, the primeval sea of hydrogen atoms (with a little helium and smidgen of lithium) was also populated with a plenitude of "free-living" quarks, the components from which hadrons are currently understood to be built, this conclusion would seem to be inescapable.

In the considerations above we have used the concept of a "force" to envisage gravity and, within the present context, this should not be a problem. The picture given by the theory of relativity replaces the notion of forces with that of distortions of a space-time continuum such that the apparent deviations from uniform straight-line motion that characterize gravitation are actually geodesics, the equivalents of straight lines, on the curved "surfaces" of these distortions. Also that masses, too, are simply distortions of this same continuum.

The picture provided by particle physics, as yet not compatible with that of relativity, represents "forces" as interactions between some classes of particles that are mediated by other kinds of particle. A view-point for which there is much evidence but is made recondite, I suspect, by our everyday notion of a particle being somehow flawed. Neither of these specialized viewpoints seem to have particular relevance to the present discussion. For which the far more intuitive approximation of force seems to be quite adequate.

# 14

# BIO-ELECTRONICS

All biological organisms are essentially fuelled by electrons having high potential energy. The engines that drive its various processes, such as the Krebs cycle, are powered by a flux of electrons from a high potential energy state to one of lower potential energy.

In overall function this process corresponds exactly to that of an electric battery, although the way that he electrons are transported are very different. So, in the interests of brevity and (I hope) clarity, I will use the positive (low energy) and negative (high energy) battery terminal metaphor here.

The source, the negative terminal, of high energy electrons varies, In plants and cyanobacteria they are acquired by the direct conversion of photons in certain bands in the red region of the visual spectrum. Animals acquire their -ve terminal electrons indirectly from those stored in plants or other animals, essentially from the high energy molecule, glucose.

The negative terminal for bacteria varies. Some examples are hydrogen, ammonium, hydrogen sulphide, sulphur , nitrite and ammonia. As might be expected, these tiny creatures require only a relatively small voltage (potential difference) to run their engines.

*But what about the positive terminal, the electron acceptor? For the motor cannot run until this connection is made. Until the circuit is completed so that electrons can flow.*

*Electron acceptors, anode, commonly used by various kinds of bacteria include the nitrate and sulphate ion, the ferric ion, sulphur and oxygen. The larger machinery typical of the eukaryotic animals requires the "full voltage" power connection as provided by use of oxygen as the positive terminal for acceptance of the electrons which have now been drained of useful energy. The oxide ion thus produced being tied up in water.*

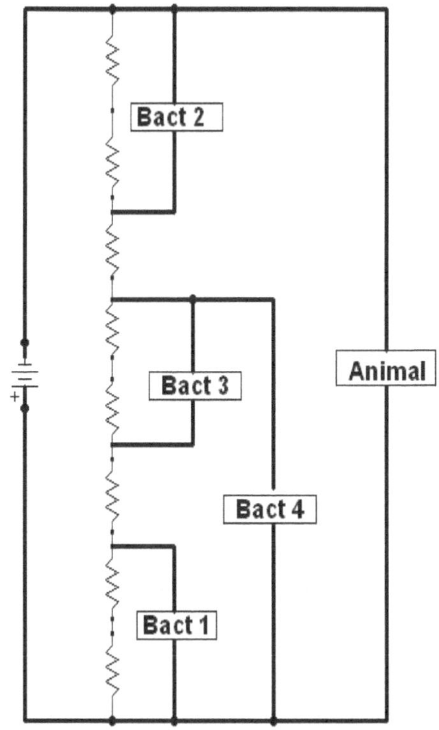


*The prokaryotes exhibit great ingenuity in ways of exploiting the energy resources of the diverse habitats that they occupy.*

*Because of their low power requirements redox pathways of lesser electrochemical potential can be used.*

*The primary battery charger for today's biosphere is, of course, the conversion of solar photons into high energy electrons in the course of photosynthesis.*

*To get energy from molecules such as sugars, bacteria must use fermentation or cellular respiration. These are processes that use the energy in the molecules to make adenosine triphosphate, (ATP). This is used for growth, movement, temperature regulation, and many other processes that require energy.*

*Fermentation, a process widely used by yeasts and bacteria, generally takes place in an anaerobic environment. By-products include ethanol, carbon dioxide, lactic acid, formic acid and acetic acid. Substrate-level phosphorylation can serve as a fast source of ATP independently of external electron acceptors and respiration. This is the case, for example, in human erythrocytes, which have no mitochondria, and in muscles during oxygen depletion.*

*Some species of bacteria use cellular respiration. This is a more complex process for which oxygen is not a prerequisite, and the energy of the organic molecules is more efficient in the creation of ATP. Carbon dioxide and water are the most common by-products.*

# 15

# IN DEFENCE OF SUPERSTITION

*"Seeing there are no signs nor fruit of religion but in man only, there is no cause to doubt but that the seed of religion is also only in man; and consisteth in some peculiar quality, or at least in some eminent degree thereof, not to be found in other living creatures.*

*And first, it is peculiar to the nature of man to be inquisitive into the causes of the events they see, some more, some less, but all men so much as to be curious in the search of the causes of their own good and evil fortune.*

*Secondly, upon the sight of anything that hath a beginning, to think also it had a cause which determined the same to begin then when it did, rather than sooner or later.*

*Thirdly, whereas there is no other felicity of beasts but the enjoying of their quotidian food, ease, and lusts; as having little or no foresight of the time to come for want of observation and memory of the order, consequence, and dependence of the things they see; man observeth how one event hath been produced by another, and remembereth in them antecedence and consequence; and when he cannot assure himself of the true causes of things (for the causes of good and evil fortune for the most part are invisible), he supposes causes of them, either such as his own fancy suggesteth, or trusteth to the authority of other men such as he thinks to be his friends and wiser than himself."*

Thus spake Thomas Hobbes. very presciently, in 1651. Today, our greater understanding of the "big picture" that has been provided by many disciplines but, in this case, by the studies of history and anthropology in particular.

Judging by evidence that is provided by the hominim fossil record, a remarkable burgeoning of the central nervous system seems to have occurred in the era of our supposed progenitor *homo erectus.* It is reasonable to assume that this neural reconfiguration corresponded to the great expansion of imagination that characterizes our species. Imagination of an order sufficient to enable and to be concurrently catalysed and prodigiously amplified by language. The synergistic triumvirate of tertiary tool use, intra-species tribal warfare and also inter-tribal socialization providing the powerful positive feed-back required to initiate and sustain this regenerative process. A high level of amplification leading to a new-found ability to form far more detailed models of the external world than those of any other existing organism. Together with mechanisms for "morphing" and conjoining such neural models. as well as the ability to import and export these imaginings in the form of language.

This far more intricate awareness of the world in which the human ape found itself inevitably leading some minds to try to make sense of it all, to tie observed phenomena together into some kind of coherent pattern. These early cogitators were what we could usefully regard as "proto-scientists". By analogy, spirits, anthropomorphic projections of the self, became variously attributed to such things as trees, rivers, other animals, the sun and the moon, lightening and thunder. There are, even now, primitive cultures that adhere to such interpretations.

With the passage of time the mythologies that embodied these myriad *ad hoc* superstitions became ever more complex and strengthened by the practice of rituals and emergence of linguistic art forms such as totems, carvings, the arrangement of large stones, body decorations and costumes.

The mythologies becoming increasingly allied to political power and maintenance of hierarchies as well as contributing to tribal identity. The condensation of larger and far more static communities that was brought about by the eventual emergence of agricultural settlements engendering the more sophisticated bodies of belief that we recognize as religions.

Despite the undeniably numerous and varied pernicious aspects of these religions, both past and present, we should be aware of the important function that superstitious beliefs have played in the evolution of collective human imagination that has engendered the sciences we have today. Because of our natural tendency towards anthropomorphic projections, primitive peoples would often ascribe human-like thought processes to such striking features as the glowing orb that would roam across the sky during the day, the clearly angry monster who would make great shoutings and hurl flashing spears at us. The diverse creatures who populated the heavens at night. Obviously there just had to be homunculi, little (or big) human-like guys behind it all. Indeed, all manner of things became endowed with "spirits" in this self-projective way that constituted the "best understanding" available. Mostly based on nonsense but, particularly, in later times, also taking on board and nurturing ideas with a more secure basis of hard evidence.

We must remember that the much of the legacy of early scientific discovery to which we owe so much today was carried out by devout followers of sundry faiths such as Judao-Christian, Islam, Brahman and Hindu. Among the multitude we find Newton, Darwin, Pascal, Ibn Sina, Maimonides, Avicenna, Nagarjuna and Ibn Hayyan. Institutions such as monasteries were important repositories for such information and sometimes encouraged experiment and discovery. A notable example being that of the Augustine monk Gregor Mendel, whose careful experiments with the hereditary of pea plants provided the basis for modern genetics. Very importantly, religious institutions perpetuated the practice of imagination storage and recovery, the skills and technologies of reading and writing.

Furthermore, they have inspired great advances in such things as architecture and also in the arts, which, although it is not generally recognized, are inextricably entwined with technology and science:

*Technologies associated with grand architecture inspired by religion*

Not to mention the significant role of religious faiths in fuelling that powerful motivator of technological evolution, warfare:

*Warfare inspired by religion stimulates advances in technology*

We have to also remember that, in those earlier times, folk had very little data to go on. Even in the fragmented polytheism of ancient Greece, where agnosticism generally met with little resistance, the curious minds of some more secular philosophers such as Aristotle and Plato were similarly hampered by the paucity of hard evidence on which to base their reasoning. However effective the information processing machine, or human mind, may be, the IT specialist's treasured mantra: GIGO, garbage in - garbage out, still applies.

Democritus seems to have done a little better than the others. By carefully studying and experimenting with minerals, plants and so forth, he increased the quality of input to the extent that some of his guesses as to he workings of the world turned out to be remarkably close to the mark. The often quoted prescient notion of the atomicity of matter which he developed and promoted seems to have been simply a lucky guess by his tutor, Leucippus. As was the Greek's expansion of the concept to the eventual condensation of these atoms into worlds. Perhaps better informed were his notions regarding the early history of human-kind. Certainly, far more solidly based on evidence were his realizations that the ratio of the volume of a cylinder to a cone of similar base and height is three.

Perhaps the most noteworthy among the ancient Greeks, however, was Eratosthenes, who, among his many other accomplishments, experimentally determined a remarkably accurate estimate of the Earth's diameter. Achieved by simply measuring the length of shadows cast by the sun from sticks placed in the ground.

ERATOSTHENES METHOD FOR DETERMINING THE SIZE OF THE EARTH

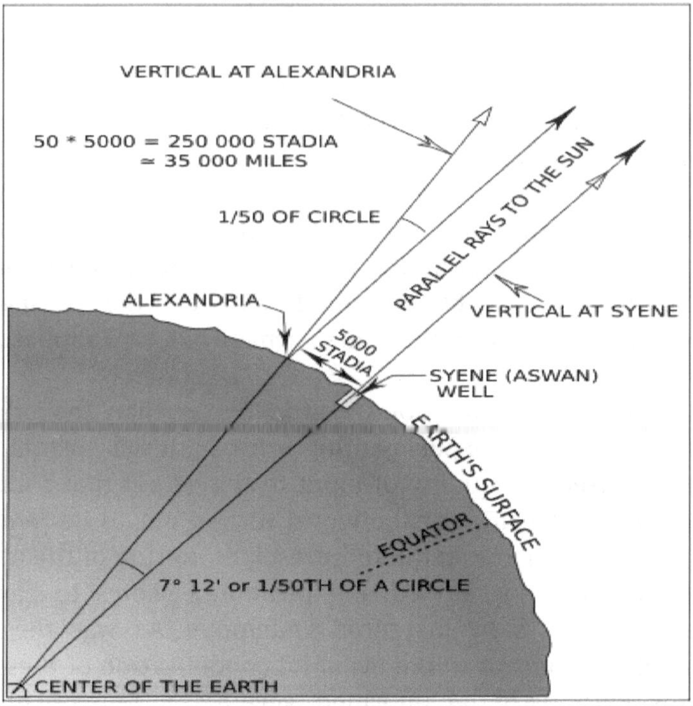


It was not until materials such as glass, copper and iron and their associated technologies became freely available that serious science could blossom. The microscope and the telescope being among the innumerable emergent technologies that provided perceptions of nature that were entirely beyond the ken of previous generations.

Of other worlds. Of a profusion of hitherto invisible, and unimagined, tiny living organisms. Of unexpectedly intricate inorganic structures. And, ultimately, the use of these crucial materials in many novel ways to provide the incomparably better comprehension of chemistry, physics and biology. The hard evidence of much of the workings of nature that we have today.

So we find that much of the central core of these belief systems, in contrast to their ritual "spin-off", is contiguous with the science of today. Historically, there is no clear dividing line. Both represent attempts to interpret natural phenomena. And, despite the bleatings of some ivory-tower academics and the unfortunate students that they manage to indoctrinate with blindly idealistic definitions and representations of science, all it really boils down to is our current "best guess". That assessment which is based on the careful analysis of information made available to our senses. Which, by virtue of the immense volume of newly gathered data that has become available to us in recent centuries, is incalculably better informed than the early efforts

So, certainly, religions have been tremendously important precursors to modern science. Essential predecessors that, in some ways, should be historically revered. But that have now become obsolete, have long since passed their use-by date. Nevertheless, even in modern times, vast numbers of folk who have been deeply indoctrinated with the religious myths and other silly superstitions such as astrology still doggedly adhere to these beliefs. Predominantly those who have had no significant exposure to the incomparably better evidenced interpretations of the workings of nature that we have today. Folk that, quite understandably, because of their circumstances, know no better. Sadly there are also many who insist on clinging to the various myths even after becoming acquainted, to a greater or lesser degree, with the new evidential models. Indeed, there are some to be found working quite competently within their own scientific fields but seem to have serious "blind spots" with regard to the "big picture".

Yet again, we find some who actually do have a good knowledge of the main body of science and can partly justify their underlying residual religious beliefs by the artifice of pointing out, very properly, the directionality of evolution and the "fine tuning" which has made biology possible. These features, they claim, indicating "design" and, by implication, the existence of a "designer". Which can then be used as an excuse to plug in their underlying religiosity in the form of a creator. Perhaps the most significant group trying to justify "intelligent design" in this way is the cult that goes under the name of "The Discovery Institute". In general, they accept nearly all the findings of modern science. Along with the general promotion of ID the members also like to push the rather bizarre notion that their reverently imagined "Creator of the Universe", their Great Intelligent Designer, has gone to all this trouble so that certain little snoutless apes (representing, we presume, the "chosen people"?) can potter around "discovering things".

Among the foremost members of the cult are Guillermo Gonzalez and Jay W. Richards, who co-authored the book *"The Privileged Planet : How our place in the cosmos is designed for discovery"*. If not for the very obvious ideological baggage with which it is laced it would be an excellent volume. Many of the main contenders for inclusion in the wide body of evidence that support the phenomenon of "fine tuning" are introduced. Some novel, some intriguing, some very plausible, some much less so.

Much is made of the fact that our moon has exactly the right combination of size and distance to subtend the same angle as the sun which it can thus eclipse. It is argued that the observation of the solar flares made possible during such an event was crucial for the development of both spectroscopy and the realization that helium was present within the sun. While a striking coincidence which does certainly give one pause for thought, it is certainly not the only route for the development of spectroscopy and for the eventual discovery of stellar helium. Or, for that matter, to significantly impede the evolution of technology.

Their arguments in favour of discarding the Copernican principle are specious. Most hard-nosed empiricists would, in any case, accept that this tool, like Occam's razor, is a well proven rule of thumb rather than a fundamental assumption. Altogether Gonzalez and Richards make a pretty good fist of underlining that our species, and the network which enables its activities, is exceptionally well adapted for the business of discovery. And, indeed, it is undeniable that we do a heck of a lot of discovering. Where they fall flat on their faces (like many others) is in the acceptance of the traditional anthropocentric concepts of "design" and "designer" The well worn intuitive notion that humans design things and that we have designers in our midst is very useful within the context of our everyday activities. Within our local networks. As also can be the simple planetary model of the atom proposed by Niels Bohr. And, say, Newtonian mechanics. But, as with the last two examples, when we get down to the nitty-gritty we find that things are not quite right. That, under closer scrutiny, the concept just does not really hold water.

Religion is founded upon the hearsay of myth rather than observation of natural phenomena. That is not to say that issues often attributed to religion: morals, ethics, the very noble aspects of human cooperation (as well as the exceedingly ignoble contributions of warfare, rape, torture, infanticide, genital mutilation and sadism) are non-existent. They are all either required components of, or spin-off from, the overall evolutionary process. Although, from this evolutionary perspective, there is no meaningful distinction to be made between properties such as "good" and "evil". Nature's machinery merely gets on with its job of generating ever greater levels of intricacy. Good and evil are meaningful only within the human context. Ethics are ***our*** concern, and ***our*** responsibility, not nature's.

Interestingly, belligerence and sociality actually turn out to be conjugates. Conflict ***between*** tribes necessarily reinforcing the need for cooperation ***within*** tribes. Cooperation, at this modern stage of human development, coming strongly to the fore. Contrary to popular belief, we find that, on close examination, the world actually appears to be becoming more peaceable. This point is well made by Steven Pinker in his 2011 book "The Better Angels of our Nature". Even so, an upsurge since its publication gives some concern. We may not be out of the violent woods even now!

Those earlier attempts to explain the world with very little data to work on were indeed the precursors of science. With the advent of such things as the telescope and microscope, wonderful new vistas were suddenly made available to us. What we call "science" today, hitherto labelled as "natural philosophy", is simply an incredibly more far-reaching and self-consistent set of explanations based upon this new and enormously greater fund of information.

Whether we like it or not, despite the many evil practices inflicted by religions and other superstitious beliefs upon humankind, they, along with our instinctive belligerence, greed and other traits that are very properly considered to be undesirable by most of us today, have played absolutely essential rôles in the evolution of technology and structure of the world in which we now find ourselves.

# 16

# THE HUMAN ANIMAL

*Views as to the part that the human species plays in the scheme of things tend to be strongly polarized.*

*The traditional beliefs derived from mythology invariably ascribe a major role for humankind. Even their deities are generally anthropomorphic projections of some kind. Even in the eyes of those cults whose deity is vaguely defined as "an intelligence", we are usually "the chosen people" The pinnacle of creation to which all life aspires.*

*At the other end of the spectrum we have a body of devout absolutist atheists who, in violent rejection of the old ideas, throw objectivity to the winds and vehemently deny that our species is anything but a concatenation of entirely random events. That we are "accidental" and there is absolutely nothing special about us or the world we inhabit.*

*The remarkable incongruity is that the latter group includes many who not only accept but are very well-versed in evolutionary biology. Who, in more rational moments, would readily agree that the filter of natural selection renders the process non-random.*

*So is our species special? Is "Human Exceptionalism" in any way justified? Certainly! Just as much as "Elephant Exceptionalism", "Amoeba Exceptionalism", "Kangaroo Exceptionalism" and "Oak tree Exceptionalism"*

There is actually another claim to fame to be made by the human, one that is shared with, among others, the flatworm. That it happens to lie on the main evolutionary trajectory that leads to the next phase of network development.

All species have their special abilities. Exquisite (and sometimes bizarre) adaptations to their respective niches.

For example:

The Star Nosed Mole

*The particular set of unique features that characterize our own species can be most succinctly be summarized as follows:*

- *Unusually complex sensorimotor innervation of the hands.*
- *Unusually complex sensorimotor innervation of the lips, tongue and larynx areas.*

*This, in fact is what makes us human. Not some vague mystical property of "intelligence", but simply being, in terms of neural architecture, all hands and gob!*

*[This representation is a sensorimotor homunculus where size is proportional to innervation.]*

What do these have in common? They are our primary communication channels.

The first enabling and, in turn enhanced by, the use of tools. The stick with which graphics can be graven in soil or sand. The pointed rock that can be used to inscribe pictures on a cave wall, The paint brush and paint, the pencil, the pen. These being inextricably entwined with the transmission of imagination as graphic language. Which having, in the course of time, become stylised as letters and words or ideograms, is today implemented by such tools as the pen and the keyboard. The concomitant sensory feedback implicit in the this manual dexterity providing an oft-unappreciated linguistic input channel with a capability approaching that of visual input. As underlined in chapter 33

The second is that enabling complex vocalizations which, of course, corresponds to the other main communication channel for the export of imagination. Together with exaption and enhancement of the auditory system as the complementary import channel. Particularly in regard to the greatly increased requirement for signal processing, of which the brain structure known as Wernicke's area plays a part.

Believe it or not, as argued in chapter 25, the presence of these two features, along with the co-evolved imagination and concomitant additional memory can account for the full gamut of unusual behaviours exhibited by this animal. Moon landings not excepted.

Note, too, that the feet are also somewhat enlarged. In fact a relatively high level of innervation of hands and feet is a notable characteristic of the primates as a whole. The human ape's particular enhancements of this feature being the amplification of those areas involved with extensive tool-use and transfer of imagination.

# 17

# THE ILLUSORY INVENTOR

The "Brilliant Mind" is a misnomer. One of the defining characteristics of humans is their tremendous variability in talents. To what extent hereditary components and those acquired by post-natal interactions with the world are responsible is not easy to determine. That both of them do contribute is certain. Even in a litter of cats brought up in the same environment, the temperaments and behaviours can be seen to diverge as the kittens develop and then eventually become adults. We can perhaps best attribute this to the amplification of tiny probabilistic environmental, genetic or cellular perturbations during development. With vertebrates, at least, the establishment of "pecking order" is very common. Even during the development of the embryo, probability plays its part. A feature which appears to be inherent in all nature's machinery. There is certainly no guaranteed causal pathway from zygote to neonate. Contingency plays its part to the extent that in this process catastrophes quite often occur along the way. They may result in abortions, stillbirths, or serious malformations. Even human identical twins are some-times found to have very different personalities and talents. The remarkable diversity that is found in human behaviour is a reflection of its inherent complexity, greatly magnified by the very high levels of feedback made possible by language and closely-related interactions with the prevailing artefacts and social structures. Increasing the variability by orders of magnitude. For example, an entirely unprecedented range of behaviours have arisen simply by the interaction of today's teenager with the mobile phone.

If we analyse the concept of the "Brilliant Mind" we find it to be essentially a hub in the evolving network of ideas. Largely a function of the individual's temporal and spatial location in the network. Being in the right place at the right time. Clearly, the achievements of Einstein, Curie or Pasteur would not have been possible had they been raised in the Pitcairn Islands. Or lived a thousand years earlier. It is not quite as simple as that, of course, because the intricate notional network is both scale-free (a property which gives it a high degree of robustness), and multidimensional, as follows.

Extrinsic requirements:

- The network must have evolved to a state where the pool of ideas contains all or most of the data required for a significant recombination to occur.

- Facilitative artefacts are available as required for acquisition of new raw data .

- The particular mind making the "breakthrough", the striking revelation, must have the time and resources required. This often equates to wealth. And usually, at least in earlier times, that massive transfusion of imagination that we call formal education.

- It must have access to the shared network by means of libraries, interest groups and, today, mass media and the Internet.

Intrinsic requirements:

- That mind must be structured in such a way that motivation towards the particular subject is sufficient. Curiosity is an important component.

- It must have a capability for making lateral associations. Thinking outside the box.

Newton's particular intrinsic qualities would have been of no avail if, like many of his contemporaries, he was working in a mill eighteen hours a day, or toiling at the coalface, or raised as a serf in Argentina, or as a Kazakh nomad. Many of whom, we would statistically expect, could well have been endowed with similar intrinsic qualities by the omnipresent machinations of chance. In striking contrast, Newton was blessed with sufficient funds and ample spare time to engage in his various ruminations, experimentations and follies. Even more significantly, he was intimately connected to a network that was abounding with ideas ready for the plucking. In the case of the calculus of variations a great deal of the groundwork had already been done by a chain of individuals traceable back through Gregory, Barrow, Hudde and Fermat. An incremental evolution of ideas having some of its early roots in the work of Archimedes. So we can dismiss the "Heroic" view of Newton's "Brilliant Mind". His achievements being better viewed as a mostly a consequence of his fortuitous location at a significant hub of the evolutionary network. Such iconoclasm may seem unkind but, in the best interests of an objective appraisal of the workings of the technosphere, it is very necessary.

Somewhat similar sentiments were expressed by Marvin Minsky in his 1982 paper beguilingly entitled "Why People Think Computers Can't" :

> *"Do outstanding minds differ from ordinary minds in any special way? I don't believe that there is anything basically different in a genius, except for having an unusual combination of abilities, none very special by itself. There must be some intense concern with some subject, but that's common enough. There also must be great proficiency in that subject; this, too, is not so rare; we call it craftsmanship. There has to be enough self-confidence to stand against the scorn of peers; alone, we call that stubbornness. And certainly, there must be common sense. As I see it, any ordinary person who can understand an ordinary conversation has already in his head most of what our heroes have."*

If we can only shake off our very natural anthropocentric biases it becomes clear that, except in a trivial everyday sense, there are no inventors, no designers. We do, of course have some discoverers, those who happen to be the right types, in the right place at the right time, who pick the low-hanging fruit. But objectively, we are forced to interpret science and technology as evolving autonomously within the collective imagination of our species.

Does the reader honestly believe that without Faraday we would have no electric motors or transformers or benzene? No mathematical understanding of the electromagnetic field if James Clerk Maxwell had never lived? Without Marie Curie we would not be aware of radium? We would have no radio without Marconi? Without Newton or Liebnitz we would not have the calculus of variations, or relativity without Einstein? The printing press had Gutenberg never lived? Or that without Steve Jobs we would not have computers with GUIs and pointing devices and other gross features not too far removed from the Apple Mac?

The ideas behind the very thoughts I am trying to express here are themselves not new. In 1922, William Ogburn and Dorothy Thomas published in a paper entitled "Are Inventions Inevitable":

> *"The dependence of an invention upon its constituent elements is a fact. The constituent elements are each in turn dependent on their constituent elements, and so on back to the ice ages and to resources of nature. But does the existence of all the constituent elements of an invention make that invention inevitable? Given the boat and the steam engine, is not the steamboat inevitable? This tendency toward the inevitability of an invention, once given the constituent parts, and the dependence of the invention on these parts, may be seen in the history of the steam engine. One sees in the interesting development of the steam engine that this invention was not dependent upon any one man and the history indicates that no one*

> *man could be expected to invent the various constituent parts as preliminary steps to making the culminating invention."*

They then go on to explain:

> *"Omitting from consideration the earlier origins of the steam engine, we may start with Rivault, who proved in 1605 by experiment that water confined in a bomb-shell and heated would explode the shell. Porta had previously described an apparatus by which the pressure of steam could be made to raise a column of water. In 161 5 de Caus constructed a machine similar to the one described by Porta. In 1630 Ramsey patented a " steam machine." This period was devoted largely to speculations as to the possibilities of steam and no further practical application was attempted until 1663, when Worcester constructed a machine similar to those of Porta and of de Caus, and used it to elevate water at Vauxhall. Hautefeuille in 1678 proposed the use of a piston in the steam engine and Huygens first applied this principle. Engines which were a decided improvement on Worcester's were now built. Great interest was aroused in their possibilities, and many minds set to work to solve the "problem of the steam engine". An important advance was made by Thomas Savery, who in 1698 patented a design for the first engine used in pumping water from mines. Improvements on Savery's engine were made by Desaguliers in 1718, by Blakely and Ridgeley in 1756, and also by Papin about this time. Thurston remarks that " at the beginning of the eighteenth century, every element of the modern steam engine had been separately invented and practically applied ".The nature of the vacuum and the method of obtaining it were known. Steam boilers capable of sustaining any desired pressure had been made. The piston had been utilized and the safety valve invented".*

George Basalla, in his most interesting book published in 1998 "The Evolution of Technology", introduced more variations on the theme. However, strongly influenced no doubt by peer pressure as well as the assumption, which is now known to be grossly inaccurate, that biological evolution has an open tree structure rather than the reticulum of the evolution of ideas, was at great pains not to imply any direct correspondence between the two evolutionary phases. Ironically, in the one passage in the book where poor George does summon up the courage to take the bull by the horns, a classic bloofer results:

> *"by the mid-1980s the home computer boom appeared to be nothing more than a short-lived, and for some computer manufacturers, expensive fad. Consumers who were expected to use these machines to maintain their financial records , educate their children, and plan for the family a future ended up by playing electronic games on them , an activity that soon lost its novelty, pleasure and excitement. As a result, a device that was initially heralded as the forerunner of a new technological era was a spectacular failure that threatened to bankrupt the firms that had invested billions of dollars on its development."*

On that lighter aside, it is time to remind ourselves that because evolutionary processes lie at the heart of innovation, this is not to say that we need to dispense with concepts of "design", "invention" and so forth for everyday human activities. Any more than with the principles of Newtonian mechanics. These are useful and very practical concepts as long as they are not used outside their domain of utility and their wider limitations are properly understood. For designing a dress, or a new kitchen layout, or a house or a complex urban development, or an airliner or a nuclear power station, the commonplace notion serves us just fine. Although, if we were to take the trouble to look closely, we will find that an evolutionary process is taking place in the background, that all such apparent creations basically stem from our pool of shared imagination rather than solely from the individual through whom they happen to be expressed.

That does not imply, though, that the outcome of your particular pet project can be said to be in any way predetermined. Free will, the ability of an organism to make choices, is, in fact, a vitally important component of the stochastic ratcheting mechanism which is characteristic of both biosphere and technosphere. It is only gross outcomes that have a high level of certainty.

It is not uncommon, or necessarily inappropriate, for us to use different perspectives for different situations. Just as we mainly engage in sex for pleasure rather than procreation. From the perspective of the average human, most of the time pleasure is the motivation. For myself, only twice in my life have my intentions corresponded to nature's function of baby-making. All the rest has been the sheer enjoyment of the delightful spin-off. Again, within our cognitive landscape, its a matter of " horses for courses."

With the transition from the relatively simple task of creating a new dress to the more complex affair of planning an urban development, the number of individual brains that is involved increases. We might very easily assume that only one brain, one imagination would be required to create a dress or even to design a kitchen. Or so it might seem using our everyday way of thinking.

But supposing we zoom in more closely to address the question of "how many humans does it take to design a kitchen? OK, so you come up with these ideas. But, hang about, where do these ideas come from? No matter how much you may pride your originality, you cannot, in all honesty, claim to have plucked them out of thin air. OK, you want an oven somewhere. Where did the concept of an oven come from? Well you picked that up from your parents when you were a child. The same applies to the other kitchen appliances, each one of which is the product of many thousands of ideas. Furthermore, you have in your mind lots of kitchen layouts which you have either directly encountered or picked up from TV shows or magazines. You have absorbed ideas regarding the pros and cons of doors and drawers. Of hinged or sliding arrangements. Of electrical and plumbing arrangements, of aesthetics. Rather than truly creating a totally original design "out of the blue" you are comparing, merging and recombining thousands of pre-existing ideas. Some learned in the "school of hard knocks", by direct experience, but, in all probability, most of them imported from the vast collective pool of shared imagination. An immense communal database which has its evolutionary roots going back two million years or so. Precipitating a self perpetuating, self accelerating process that thrives on the transfer of imagination which we know as language.

Whenever we look closely enough, we find no firm evidence of human original design. We find that invention is an illusion. An illusion deeply reinforced by myth, human psychological needs, and human emotions. Myths continually reinforced by legend, by marketing and, not infrequently, by some of the "inventors" themselves. For nearly all those of us who enjoy exercising our ingenuity are proud of the novel syntheses of ideas that arise in our minds. For those so inclined, a great kick is to be had from knocking up a new gadget or an original work of art, or a food recipe, or computer app. And if it is recognized or appreciated or earns some dosh, so much the better.

But, hey, surely Edison invented the electric light bulb? Well that particular myth is actually on its last legs by now. Quite a lot of folk are now well aware that the first carbon filament bulbs sold by Edison were a rip-off of those devised previously by the well heeled Englishman Joseph Swan. Edison, although he instituted and ran a great laboratory, seems, at least in his later years, to have excelled as more of a crooked businessman than as a synthesizer of gadgets. But even shady tycoons have their place in techno-evolution, and the implementation and marketing of an electricity reticulation system that could compete with that already established for gas lighting was Edison's real synthesis.

So Swan was the real inventor of the incandescent electric light bulb? No! Over a dozen others had independently been fooling around with the idea of enclosing conductive filaments within evacuated glass bulbs well before then, with varying degrees of success. All picking the low hanging fruit. Recombining and experimenting with the new idea complexes that had recently evolved and which were manifested in such diverse things as metallic wires and carbon filaments, plentiful supplies of electricity, vacuum technology and glass working. As well as the fast-growing body of ideas representing physics and chemistry which were there for those fortunate enough to have taken the mass transfusion of formal education or to have had the opportunity to import from the literature.

But in fairness to that "great inventor", Thomas Edison, maybe the light bulb was a poorly chosen example of his genius. After all, was he not truly the first to record sound? Well, no, as it happens, for it is the case that the Frenchman Édouard-Léon Scott de Martinville had recorded sound when Edison was a ten year old whipper-snapper. This was a "write only" device using a soot-coated motor driven cylinder in conjunction with a system of levers that was connected at one end to a membrane sufficiently flexible to respond to sound and at the other end a stylus that would scratch off the soot as the drum rotated. This, in turn, was simply a particular application of the already well established kymograph, a device which was used to record all manner of mechanical perturbations, after the idea synthesis attributed to Carl Ludwig around 1840, and that, in turn had its own evolutionary roots. The kymograph was still widely used until mostly ousted by the oscilloscope in middle of the 20th century. (I still have fond memories of coating kymograph rolls with soot using a benzene burner as a teen-age lab technician working in the Pharmacology department of the Royal College of Surgeons of England. Happy days!

*The Kymograph*

A read-write disc based sound recording/playback device had been subsequently proposed by the poet Charles Cros, a full ten years before Edison's version, and Emile Berliner pipped him to the post in 1887 by building and patenting a working device along Cros's lines. The sound amplification device used by both systems, the horn, having evolved many centuries before, perhaps inspired by the conch shell, for use in musical instruments. So we see that, by this time, the great shared pool of human imagination was saturated with ideas related to sound recording and reproduction as well as materials and component gizmos. All ready to "crystallize out" to a marketable consumer device. And what more likely place to seed this inevitable development than the newly constructed and superbly staffed Edison Laboratories research complex. Funded by the fortune generated by his successful marketing of an improved version of the already well established stock ticker.Unlike many of the fabled "Heroes of Invention" Edison was not one of those "born with a silver spoon in his mouth". His early years of hustling newspapers and confectionery on trains exposed him to the idea pool of commerce, most likely setting the scene for his subsequent success in business. His hobbies of fooling around with gadgets, no doubt inspired by boy's books dedicated to such things that were still abundant even in the times of my own youth.

*Some stages in the of evolution of the gramophone*

*Morse Key*

*Regina Music-box*

*Conch shell*

*EdisonPhonograph*

*Berline Gramophone*

And he also enjoyed experimenting with chemistry, living before this chemophobic age where kids (or BSc students, for that matter) are no longer allowed to interact with all these dangerous nasty, toxic, smelly and sometimes explosive compounds. He thus had plenty of valuable exposure to the scientific idea pool! Enhanced by his visits to the Cooper Union. The latter, a free education institution was set up by one Peter Cooper, himself a curiosity freak, prolific synthesizer of ideas and entrepreneur who had become very wealthy. And whose history, to which Edison was indubitably exposed, itself providing evolutionary links to the youth's mind, even further inspiring his inherent commercial interests. So it would come about that, within the well resourced team in his new laboratories, an idea evolved which was no doubt provoked by fooling around with one of the new wonder materials, tinfoil. (Aluminium was still a precious metal at that time and would not be available in the form of a foil until many decades later). In association with the recollection of the uncannily speech-like sounds produced by the telegraph receiver, with which he was very familiar, the conditions were certainly most auspicious for the next evolutionary step in the reversible recording of sound to take place. The phonograph itself was not a great commercial success, for the market lead was taken back. Emil Berliner, drawing upon the lower cost initiative of Edison to make some improvements to his own previously patented Gramophone, stole a march on Edison and so wound up with the last laugh.

The reader might by now be coming to the conclusion that I have a great contempt for Thomas Edison. Nothing could be further from the truth, I like his style. His life story is absolutely fascinating. His achievements were driven by commerce rather than warfare. A trend which, as detailed by Stephen Pinker in "The Better Angels of our Nature", seems to be on the up. Also I feel a bond inasmuch as I, too, love to tinker with ideas, with chemistry and with gadgets. But, when push comes to shove, an inventor he ain't! Simply because there really ***are*** no true inventors. Only evolutionary hubs through which, because of its structure at any point in time, there happens to be an unusually high flux of ideas within the network. Is this a jaundiced way of looking at things? I think not. Because I may happen to recognise the broad evolutionary processes that have led to the structures of a rose, and also being aware of the biological reasons for the effect it has on me, that does not make it any less beautiful. A sentiment very nicely expressed by Richard Feynman:

> *"I have a friend who's an artist and has sometimes taken a view which I don't agree with very well. He'll hold up a flower and say "look how beautiful it is," and I'll agree. Then he says "I as an artist can see how beautiful this is but you as a scientist take this all apart and it becomes a dull thing," and I think that he's kind of nutty. First of all, the beauty that he sees is available to other people and to me too, I believe. Although I may not be quite as refined aesthetically as he is ... I can appreciate the beauty of a flower. At the same time, I see much more about the flower than he sees. I could imagine the cells in there, the complicated actions inside, which also have a beauty. I mean it's not just beauty at this dimension, at one centimetre; there's also beauty at smaller dimensions, the inner structure, also the processes. The fact that the colours in the flower evolved in order to attract insects to pollinate it is interesting; it means that insects can see the colour. It adds a question: does this aesthetic sense also exist in the lower forms? Why is it aesthetic? All kinds of interesting questions which the science knowledge only adds to the excitement, the mystery and the awe of a flower. It only adds. I don't understand how it subtracts."*

There you have it, For those who are prepared to temporarily take off the blinkers we need for everyday life, there is nothing so awe-inspiring as this wonderfully intricate evolutionary network of things and ideas. The ideas I am promoting in this book, as I once more emphasise, are not new. Many have been around for a very long time. My only "original" contributions really just involve an adjustment of perspective from the anthropocentric to the objective. And also an integration of the separate well accepted evolutionary processes into a continuum that, in light of present knowledge, can be reasonably projected a little way into the future. This is a transition which we naturally tend to find distasteful. But because of the present state of the network, because these ideas are now ripe for the plucking, they are starting to become rather obvious, so many others will be already thinking along similar lines. Duplication of this kind is ubiquitous and, in itself, provides strong evidence for the evolution of technology.

I have personal experience of this. In the course of my very chequered career, which includes but is by no means limited to being various kinds of chemist, as well as sound technician, security guard, purchasing officer, manufacturer, itinerant, and so forth, I have from time to time come up with gadgets and processes that might be worth patenting. Nothing earth-shattering, you understand. Just bibs and bobs. A few I took to provisional patent stage before running out of enthusiasm (as you may have guessed, I suffer from "butterfly mind" syndrome). Some of these just got to be notes for future reference and hoarded. Doing anything with patents used to be either expensive (using an agent) or a major hassle (DIY). Taking long trips to the patent office, wading through microfiches. A major pain in the butt. But not long after the World Wide Web was at last up and running it suddenly all became very easy and cheap. So, just as an amusing diversion, I dug out my documents and searched accordingly. To find that, with very few exceptions, patents had popped up for the same "invention" around the same time. Interestingly, one of the criteria for rejection (or contesting) a patent is "obviousness from the state of the art". Were this rigorously enforced I strongly suspect we might have no patents approved at all! I guess it depends on how obvious and to whom. Clearly it is obvious to the applicant himself or herself.

While on a personal note, I have always been surprised to find that when I set out to make a little gizmo to serve some useful purpose around the home or workplace, whatever it may be often seems to take over and devise itself. (Many writers find this, too, the characters in a novel, for instance, seeming to take over and assemble the plot). I may sit down and make a provisional sketch before I start, but when I get into my workshop I am often inspired by various bits of wood, metal, plastic and the miscellaneous junk that I have lying around. One thing seems to suggest another and, more often than not, the result does the job required of it but bears absolutely no resemblance to the sketch with which I started. It seems to be quite common among folk that fiddle and twiddle. Sometimes it is referred to as "eyeball design". A phenomenon which perhaps motivated Edison's requirement that his mouth-watering Menlo park "ideas factory" be equipped with:

> *"a stock of almost every conceivable material...eight thousand kinds of chemicals, every kind of screw made, every size of needle, every kind of cord or wire, hair of humans, horses, hogs, cows, rabbits goats, minx, camels...silk in every texture, cocoons, various kinds of hoofs, shark's teeth, deer horns, tortoise shell...cork, resin, varnish and oil, ostrich feathers, a peacock's tail, jet, amber, rubber, all ores and minerals..." and much, much, more!"*

So, if we accept, as it seems we must, that, except as a trivial but useful everyday fiction, a little white lie, there are really no inventors or designers. That, truly, technology evolves by the accumulation of ideas within our collective imagination, both from the environment and from the continuous shuffling and recombination into more intricate complexes, which, to a large extent, takes place in individual minds. We also detect a certain inevitability in the overall process inasmuch as niches that arise in the technospheric network are filled. Filled by such things as sailing ships, steam engines, electric motors, radio sets, the camera, television sets, the desktop computer, aircraft and so forth.

So does this mean that all the artefacts generated by us are predetermined, that techno-evolution "runs on rails"? No, for this is not the way that we observe nature to work. Its machinery clearly does not run on rails but rather by a probabilistic exploration of the growing network. And the underlying properties of that network are such that niches arise that are readily filled by selection between certain members of the vast set of idea complexes that are generated by "mutation" Those of a mathematical bent might like to envisage these ever-occurring new niches as "attractors.

The palaeontologist, Stephen Jay Gould famously proposed a "thought experiment" regarding biological evolution. He called this speculation "replaying life's tape.":

> *"You press the rewind button and, making sure you thoroughly erase everything that actually happened, go back to any time and place in the past.... Then let the tape run again and see if the repetition looks at all like the original"*

His expectation was that:

> *"Any replay of the tape would lead evolution down a pathway radically different from the road actually taken."*

Since "replaying life's tape" or, in our present case, technology's tape, is out of the question I will reset it in a form which is at least conceivably do-able and substitutes technology for biology. Let's suppose, for the sake of argument, an earth-like planet on which a set of human equivalents had evolved and achieved a level of technology quite similar to that of our own. Among the billions of artefacts they produced, would they include an Eiffel Tower? Would they have a replica of the Vatican, complete with a reproduction of the Sistine Chapel and its ornate ceiling? Van Gogh's Starry Night? Big Ben? Beethoven's Fifth Symphony? We would have to consider the recurrence of even one of these as having a vanishingly small probability. But, on the other hand, we can be pretty damn sure that they would have the same engineering materials and basic geometries that make the Eiffel Tower feasible. They would indubitably have elaborately ornamented religious architectural works, and painters and clock towers and musical compositions. And, almost certainly, sailing ships, electric motors, aircraft, electricity distribution pylons of similar shape and so on. So how do we account for the disparity between these groups? Firstly we can say that duplication is improbable in the details. Also we can regard a very large proportion of the artefacts (which include social constructs like Companies and religious cults) as contingent "spin-off" from the main evolutionary process. Inevitability being constrained to items at the core of the process, those vital to the functioning of nature's probabilistic but ultimately ineluctable machinery. We shall return to Gould in a later chapter.

# 18

# COMPLEDEPASCENCY

## Google meets the Burgess Shales

Many a long year ago, during the Luftwaffe's blitzkrieg, my mother and I escaped from London. Passengers in a lorry which my enlisted father had arranged. A night time trip, of necessity by moonlight, since civilian vehicle lights were forbidden. Our destination was Bumpers, just outside the beautiful and then unspoiled village of Ilmer, nestled close to the gently rolling Chiltern hills. Unknown to my toddler self and to almost all adults as well, only a short distance over those hills sat the world's first programmable electronic digital computer, Colossus. Bletchley Park's secret code decryption machine. My first fairly near encounter.

Tinkering with electronic gadgets and inspired by magazines such as "Wireless World" "Popular Electronics" and "Practical Wireless", I managed to keep track of the main developments in computation. My teenage kid brother, John, on leaving school, became a programmer for an insurance Company. Thereafter he, as well as some of my friends, drifted through such central IT hubs as the Lyon's Teashops group, (Leo) and London's prestigious Oxford Street department store, Selfridge's.

So although at that time largely occupied with chasing girls, motorcycling, boning up on philosophy, chemistry and physiology, pub crawls, frequenting jazz clubs and concert halls, hitch-hiking in Europe, swimming, tennis, and so forth,

I still managed to keep a general eye on mainframe computational adventures.

After marrying, I was for a while involved in some physical chemistry research which involved some serious calculation. Even using an, as then, state of the art Facit mechanical desk-top calculator, this was quite time-consuming. With my wife and baby daughter I departed for New Zealand just before a new computer terminal was commissioned for my work-station. Another close shave.

So it was not until the release of the Commodore 64 that I finally started to get my hands dirty. Burning the midnight oils learning to code in the rather primitive Commodore Basic so as to produce a program that would handle most aspects of the manufacturing business that I was running at that time. Rather amazingly, it worked well, although de-bugging my ab initio "spaghetti Basic" code could be a nightmare.

Those were heady days, for the entirely new niche of personal computing had emerged. There was an enormous diversity of quite distinct species of computers, all proliferating to fill this new habitat. As with biological species, they used the fundamental body plan inherited from mainframes but implementations varied greatly. Some of the long-forgotten names, much-vaunted in their time, were Acorn, BBC, Elektor, Quest, Commodore, Netronics, Nascom, Heathkit, SCELBI, Amstrad, Atari, BBC, TRS-80, Sinclair, ZX80, Altair 8800. Many of these, in turn, implemented in a variety of configurations, home-builts and pre-manufactured.

Many of these bushy early efforts, as with nature's animal experiments during the "Cambrian explosion", soon fell by the wayside, and now only a few fossils of these machines are to be found in museums.

As years passed, the number of species thinned out until only two lineages persisted to any great extent. Apple, traceable from the early diversion and the PC, which had a slightly different genealogy, having been added to the scene a little later by IBM and thus more directly derived from mainframes. The philosophy behind this latter system being that of "open architecture" a rather clever scheme allowing a variety of software to be used, at the same time constraining to some extent the inevitable attempts to close-copy the hardware and operating system, so as to at least maintain the integrity of the species.

A couple of decades after my initial skirmish with the C64 we had passed through various other stages of completedepascency, including the many communicative and gaming species that evolved within the naked Internet before the emergence of a new niche which was labelled "The World Wide Web" and its initial occupation with a goodly number of proto-browsers of which Mozaic was among the first. An inevitable development within this rapidly expanding new environment was the search engine.

In truly completedepascent fashion they rapidly diversified. Freed from the inconvenience of FTPing around the net we found Gopher, then Archie, then Veronica, then Jughead. Within a few years this habitat would also come to contain, among others, Ask Jeeves, Dogpile, W3catalog, Allweb, Lycos, Webcrawler, Yahoo, Jumpstation, Go.com. Infoseek, Magellan, Daum, HotBot, SAPO and Northern Light.

Sometime in 1998 I had grown tired of drifting around search engines, many of which had become cluttered and some not very efficient. Being a fan of P G Wodehouse, Ask Jeeves held me in its spell for awhile, but I had mostly settled on Altavista.

*On a whim I browsed around to see what was new and found a start-up that used an interesting new algorithm based on identifying network hubs so as order the search return sequence. Better still, this search engine had a beautifully uncluttered interface. The coloured letters of its name reminded me of the toys that I had played with in my first year at school. And the name itself was a nice pun on that used for $100^{100}$.*

*So I fell in love. As over the years, have most folk. So much so that the noun Google, along with other parts of speech has become firmly embedded in our language. Interestingly, in this instance, although Google has achieved supreme predominance, quite a few of the early species still survive. Which goes to show that network dynamics depend not on compledepascency alone.*

*Moving on a few more years in this personal saga I became aware of the E-bay phenomenon. Already established in the USA and England its tentacles had not yet spread to New Zealand. A nice easy niche was there for the filling. For a while, every Tom, Dick and Harry with an entrepreneurial leaning and the rudiments of IT was putting up a trading site. Ready-to-go software was even on offer for those lacking the IT background.*

*This was a great burgeoning of species reminiscent of that revealed by the Burgess Shale. And in contrast to the search engine example, the effective survival of only one significant New Zealand site was to ensue: that of TradeMe. I was watching the scene carefully at its inception. There were all manner of trading sites trying to get a foot-hold. Some were plain silly, some cumbersome, some with commissions too high to attract clients, some too low to be economically feasible.*

*There also some who were sensibly set up, with little to choose between them. Sam Morgan, whose site was in this category, got lucky. In evolutionary situations of this particular kind, in addition to completedepascency, there is a powerful positive feedback effect in action. Folk naturally gravitate towards a site where their chances of a sale or bargain are greatest. That pretty well equates to a site used by the greatest number of clients. Given a situation starting with sites of equal merit a small perturbation in the system such that one has slightly more clients will start an avalanche such that there can be only one significant survivor. Winner takes all!*

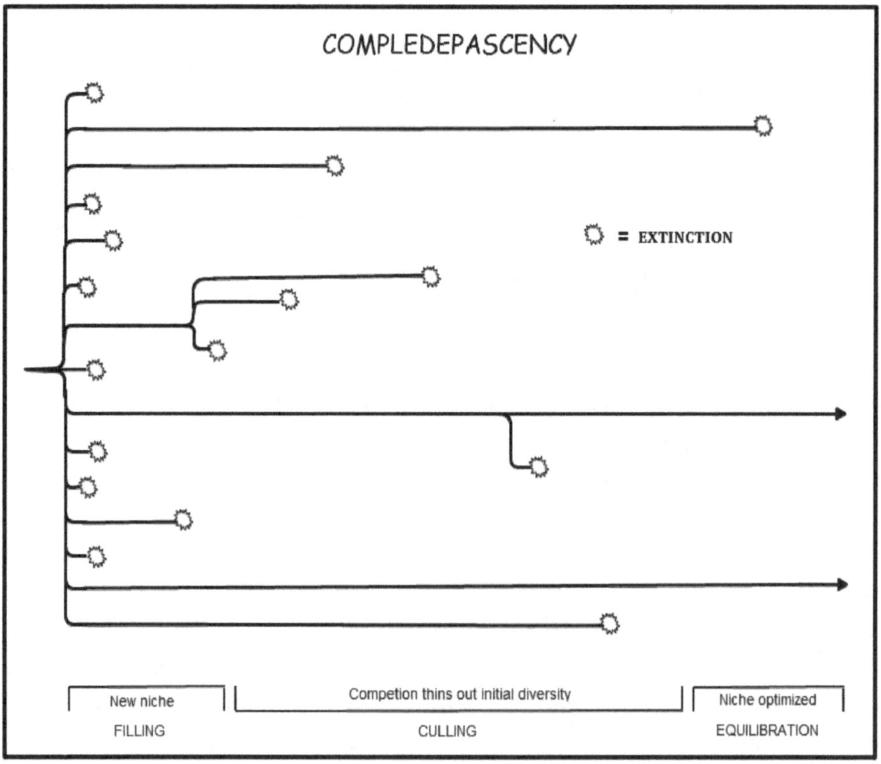


*We will return to compledepascency in its biological context in chapter 11*

# 19

# STRUCTURED STOCHASTIC SERENDIPITY

## (The stochastic ratchet)

There is always some degree of inertia in the adoption of a paradigm-shift. But old prejudices are crumbling and there is now more widespread tacit, if not overt recognition of the fact that stochastic inputs can (and do) drive the directionality of "life" processes. An overall directionality that is clearly observable in the biological and other evolutionary processes which, from our traditional human perspective, we interpret as "purpose"

Jose Ignacio Pascual et al of Freie Universität Berlin have also helped fan the flame of this better understanding by recently demonstrating that the stochastic motion of a vibrating hydrogen molecule. can be harnessed to progressively move a mechanical cantilever. And in "Life's Ratchet", Peter M. Hoffmann raises the spectre of Maxwell's demon by making a case for the extraction of energy from the random molecular motion of a liquid in thermal equilibrium by the ratcheting effect enabled by biomolecules.

I find the interpretation of these phenomena as deriving ***energy*** from purely stochastic processes to be seriously flawed. There is, however little doubt that ***directionality*** can be thus derived. After all, that is exactly the situation with a macroscopic ratchet. A ***rectifier*** of energy that drives directionality. Certainly not a ***generator*** of energy. In the cases of both the hydrogen molecule and of the cellular machinery, the system requires an input of energy to effect the rectification. In the latter case, high energyelectrons must be fed into the system to achieve the synthesis of ATP to ADP or nothing happens. Electrons being required to generate the gradient that drives the proton pump.

There is here no conflict with thermodynamics because the directional order thus produced by molecular motors is, as always, at the expense of an increase in entropy. In this case ultimately in the form of red-shifted heat.

The interpretation of natural processes of this kind as a ratcheting mechanism happens to be my own preferred analogy. A rectification of randomness which perhaps operates at all scales. It does however, require some kind of underlying structure inherent in the fabric of our universe with which the stochastic component interacts. And it is this which can provide a rational basis for the contentious issue of fine tuning of the universe for "life".

SOME EXAMPLES OF RATCHETS

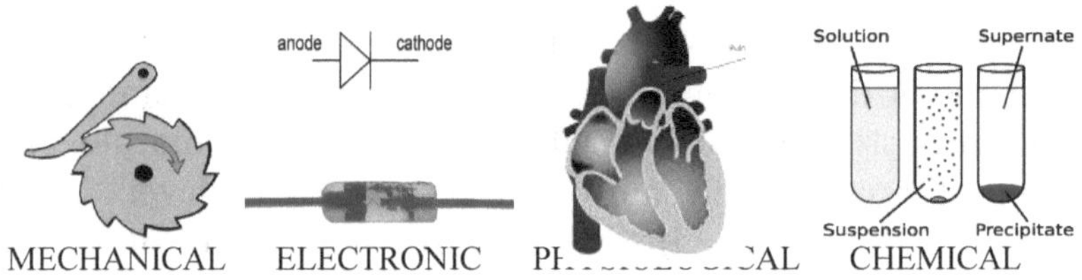

I have been taxing myself with the tricky task of finding some alternative to the oft used term "fine tuning", which is in some ways inappropriate, and most certainly misleading, as it held by many to imply the existence of a "fine tuner". Other terms I have considered include "Universal Ratchet", "Stochastic Ratchet" and the oxymoronic "Stochastic Certainty" or even "Stochastically Structured Serendipity".

The last, SSS, implies that our universe is inherently structured for evolution, that it has an underlying evolutionary mechanism. As well evidenced by its indisputable directionality and the innumerable instances of the prevailing conditions and/or entities being "just right" so as to enable the continuance of the process. In the biological phase, for instance, mutation is, according to our best understanding, mostly random. Other, more important ways in which genetic diversity arises, such as gene transfer, cell fusion and sexual shuffling, also have significant random components. The observed directionality of biological evolution is almost entirely attributable to the subtractive action of natural selection. Whereby variations that induce phenotypes which are optimally configured in such ways as to enable reproductive success in the prevailing environments to which they are subjected are necessarily favoured and tend to proliferate. In the case of biology and technology, at least, natural selection seems to be identifiable as the ratchet. It must be remembered that these environments themselves are integrated into the network as a whole, this, too, being perpetually in a state of dynamic flux. This provides a satisfactory explanation for the observed phenomenon of "punctuated" evolution. Sometimes biological evolution has to wait for the environment to "catch up", if you will excuse the metaphor. To use a very familiar example, the emergence of large multicellular organisms was unable to occur until after atmospheric oxygen levels became sufficiently high.

The evidence for "fine tuning" (SSS) and directionality of our universe for the overall "life" process is certainly not limited to the much vaunted physical parameters. Geology, biology and particularly chemistry provide numerous examples of "just right" prevailing conditions and/or materials that enable and, indeed, make virtually inevitable, the strong directionality we observe in evolutionary processes. All indications are that the observed universe has an underlying structure such that stochastic inputs ratchet evolutionary processes forward. Not only bio-evolution but also the other aspects of nature's machinery, from the formation of the chemical elements in stars and supernovae right through to the evolution of technology within the collective imagination of our species. All of which exhibit strong directionality. A pervasive and persistent pattern which can usefully be extrapolated to predict the imminent emergence of a new, non-biological predominant cognitive entity on this planet from what is now the Internet.

The presence of such a pattern of increasing intricacy is unequivocal. It is there for all to see. It is clear in the stars, in supernovae, the minerals near the surface of our planet, in biology and, right now, in the exponential development of technology. Serious interpretation of the phenomenon, however, is subject to much contention. So, before expanding upon the discussion of SSS we will briefly consider the main alternative approaches which are abroad.

- The "stick your head in the sand and pretend its not there" approach. Incredibly, despite the mountain of evidence, there are staunch members of the paradigm-bound old school who even now persistently deny evolutionary directionality. These include divers prominent academic luminaries who are nominally involved in the sciences. In addition to, as might be expected, many of the hard-core religionists.

- The latter-day deist, often operating under the cloak of "intelligent design", who happily accepts the observed directionality and even the validity of evolutionary science but nevertheless, with no real evidential support, insists upon plugging a creator into some primeval stage of nature's machinery.

This latter option cannot really be faulted except in terms of parsimony. It is common practice in science to take the simplest solution that is consistent with observed facts as the preferred explanation. But that practice, although extraordinarily useful, is not infallible. Rather, the main objection to ID hypotheses is that there is absolutely no empirical evidence to suggest a creator.

The only straw at which ID proponents can sensibly clutch is that of "design". A concept that, although very dear to our hearts, on careful inspection turns out to be illusory.

- Then we have those, often theoretical physicists with platonic tendencies, who would assure us that our home universe is but many of a multiplicity, so great in number and so widely varied in their fundamental constants that at least one of them had to be right for life.

Just as with the ID cult, such a possibility cannot be logically disavowed. And in exactly the same way as creationists invoke a deity, a whole heaping helping of complication is introduced with zero evidence and not even a possibility of real testability on the horizon. As also is the case with Everett's "Many Worlds" interpretation of quantum weirdness.

- There is also the inescapable possibility expressed nicely by Isaac Asimov (I haven't been able to find the quote) to the effect that the pattern we observe may be but an eddy in a sea of entropy.

If ,this is the case, and we, of course, cannot be absolutely certain that it is not, then the eddy is a remarkably wide-spread and long-lasting one.

- Finally, we have those who point out that the human imagination can find patterns in almost anything.

This is very true. On a nice summer's day we are very adept at finding all manner of animals, human heads and so forth among all those puffy cumulus clouds. On the other hand, illusory impressions such as this soon fall apart on closer scrutiny. We need to be careful to separate the wheat from the chaff. However, at the same time, we have to bear in mind that the ***whole*** of science is dependent of the interpretation of patterns. The periodic table and the "flavours" of quarks are particularly striking examples. No empiricist worth his salt can choose to simply ignore such oft-occuring and ubiquitous examples of "coincidence" and dismiss them as such. And so it is that we really do have to face up to the reality that, time after time, in all the evolutionary processes we find that pathways for growth of the network of intricacy are not arbitrary, but are constrained by the requirement for a set of conditions to be "just right". One of the earliest to clearly identify this proposition was the biochemist Lawrence Henderson. His book "The Fitness of the Environment", was published in 1913,begins thus:

> *"Darwinian fitness is compounded of a mutual relationship between the organism and the environment. Of this, fitness of environment is quite as essential a component as the fitness which arises in the process of organic evolution and in fundamental characteristics the actual environment is the fittest possible abode of life. Such is the thesis which the present volume seeks to establish. This is not a novel hypothesis. In rudimentary form it has already a long history behind it, and it was a familiar doctrine in the early nineteenth century. It presents itself anew as a result of the recent growth of the science of physical chemistry"*

After a very exhaustive and detailed discussion of the "just right" properties of hydrogen, carbon and oxygen he very convincingly concludes that:

> *"The fitness of the environment results from characteristics which constitute a series of maxima — unique or nearly unique properties of water, carbonic acid, the compounds of carbon, hydrogen, and oxygen and the ocean — so numerous, so varied, so nearly complete among all things which are concerned in the problem that together they form certainly the greatest possible fitness. No other*

> *environment consisting of primary constituents made up of other known elements, or lacking water and carbonic acid, could possess a like number of fit characteristics or such highly fit characteristics, or in any manner such great fitness to promote complexity, durability, and active metabolism in the organic mechanism which we call life."*

Since that time, however, advances have been made in a quite different field, namely that of astrophysics. In development of the understanding of the detailed processes involved in stellar nucleosynthesis, specifically the steps leading to the generation of carbon from three helium 4 nuclei, the team led by Fred Hoyle predicted that the formation of this element in sufficient abundance to allow of biology depends critically upon the existence of an excited state of carbon-12 that has an energy level 379 K above that of the three helium nuclei from which it is formed. The experimental verification of this prediction led Hoyle, not usually of a theistic disposition but nevertheless falling into the usual anthropic trap, to remark:

> *"Would you not say to yourself, 'Some super-calculating intellect must have designed the properties of the carbon atom, otherwise the chance of my finding such an atom through the blind forces of nature would be utterly minuscule. A common sense interpretation of the facts suggests that a super-intellect has monkeyed with physics, as well as with chemistry and biology, and that there are no blind forces worth speaking about in nature. The numbers one calculates from the facts seem to me so overwhelming as to put this conclusion almost beyond question."*

Dean Lee of North Carolina State University, having recently collaborated with a team of German physicists to further examine the process concludes:

> *"The Hoyle state of carbon is key....If the Hoyle state energy was at 479 keV or more above the three alpha particles,*

> *then the amount of carbon produced would be too low for carbon-based life. The same holds true for oxygen....If the Hoyle state energy were instead within 279 keV of the three alphas, then there would be plenty of carbon. But the stars would burn their helium into carbon much earlier in their life cycle. As a consequence, the stars would not be hot enough to produce sufficient oxygen for life. ... more than a 2 or 3 percent change in the light quark mass would lead to problems with the abundance of either carbon or oxygen."*

Although the factual basis thus appears to be sound, the opinions regarding attribution of SSS are highly polarized. The committed creationist triumphantly wields Hoyle's unfortunate remark while those at the other extreme of the theist-atheist ideological spectrum attempt to wriggle out of the superficially anthropic implications by improbable adjustments of the given data. Victor Stenger, for instance, argues that simultaneous adjustment of other parameters can be made to provide alternate pathways. Bitter controversy continues between these camps. Both of whom completely fail to realize that SSS is ***not*** indicative of "design".

There are many features of our salubrious location which are often put forward as good examples of SSS. Some of these, although remarkably opportune, can be more reasonably explained on a probabilistic basis, given the very large number of potential sites that present observations suggest to be available.

These offerings include:

- Being in the "habitable range" of our sun.
- Being in the right mass (and hence gravity) range. Much smaller giving heat balance problems, less atmosphere, precipitous mountains. Much larger, a flat terrain with only temporary land masses.
- Having a protective giant planet corresponding to Jupiter whose gravity (deviates) so as to reduce the likelihood of major collisions.

- Having a moon of just the right size and distance to neatly eclipse the parent star.
- Being in a hospitable region of the galaxy.

Although requirements such these are interesting in their own right, they in no way thus qualify for inclusion in the far more fundamental and universal category of instances of SSS that are exemplified by the special properties of such things as carbon, oxygen, hydrogen, nitrogen, phosphorous, manganese, molybdenum, calcium, silicon, copper, iron. Just a few of the many entities that, even if only one of them was not "just right" would have scuppered the evolution of the amazing networks of intricacy of both biology and technology that are so very evident today. These examples will be discussed in chapter 29.

# 20

# BEING AWARE

*It is sad to see scientists who should know better sounding off metaphysically on an issue that, in the light of modern knowledge, is entirely explicable!*

*The underlying mechanisms of the central nervous system are, of course, still far beyond our ken. Even for the workings of a single neuron, or, for that matter of any other kind of cell, our knowledge is certainly incomplete.*

*However, we do not need to be an expert in mechanical engineering to understand the overall function of a car, or even to drive it. No more than we need to understand the vast range of chemistries of its component materials.*

*There is no "mystery of the automobile"*

*Similary, if we simply restrain our instinctive self-referential departures from objectivity, we find that there is now absolutely no mystery about the basic nature and function of consciousnesss.*

The mystical notions that have arisen in the past (and often still do!) are purely illusory, an inevitable result of approaching the issue by introspection.

This, of course, was the only option available to earlier philosophers, and many still have trouble escaping from that trap with its inevitable recursive loops.

Today, although the details of the nervous system function of ourselves or other animals are still quite intractable, we have sufficient information to give us a rough idea of the gross workings of these systems.

From evolutionary considerations, we can also now see how the essentially navigational function that we like to call "consciousness", "self-awareness" "sense of agency" and so forth is bound to arise.

We can see, as well, that the incomparably higher level of consciousness exhibited by our own species compared with that of other organisms (even the higher mammals and birds) is simply a function of our much higher degree of interaction with our environment and, by virtue of language, with each other.

Furthermore, from another discipline, we now have an excellent understanding of the workings of functionally analogous computational systems. With these new tools at our disposal we can now view the phenomenon in a truly objective way. And then the hocus-pocus surrounding this issue vanishes. Be aware of chapter 27.

# 12

# EVOLUTION OF BIOLOGY

## Evolutionary trees

One of the earliest tree cartographers was the morphologist Ernst Haekal. His quite elaborate representations fell into disfavour for two reasons  Firstly, they tend to represent humankind as "the pinnacle of evolution, an anthropocentric view that was common in those times and which, sadly, is still alive and well and living in some secular circles. (In religious cults humans, or the devotees, are invariably "the chosen people"). The second misleading quality in Haekel's beautiful and topologically acceptable tree illustrations is that the vertical axis does not properly represent time .

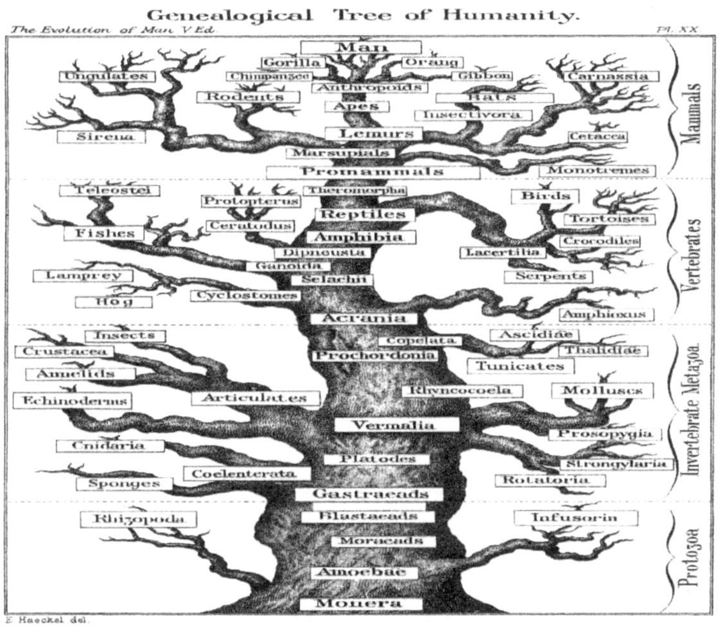

For instance, the apes appear to not be contemporaneous with our species, let alone the vastly successful family of arthropods. And the prokaryotes together with their presumed retrograde derivatives, the viruses, alone account for most of today's biomass, so in terms of sheer weight of numbers the "inverted cone" of Haekel's depictions are certainly misleading. This seems to be the main basis for the rejection of such diagrams by the followers of Gould, who fail to recognize that these representations are topological rather than temporal with respect to biomass. An alternative approach is to use "evolutionary distance" rather than time or simple topology for the construction of the tree. Whereby the number of evolutionary steps (not easy to define) is used as the independent variable.

The greatest substantive weakness of all traditional tree diagrams, however, is that used very improperly by Basalla and many others to "prove" a fundamental distinction between the evolution of biology and that of technology. It relies on the belief that branches of the biological tree are always divergent. A view which, in the last half century, has been resoundingly falsified. The evidence all pointing to the fact that the biological evolutionary tree, like its technological counterpart is not a "tree" at all. Its topology, as indicated in the illustration below, attributed to Ford Doolittle, is instead that of a network. This, as I hope the reader may have by now become aware, itself being a great simplification. For this is merely a network within a much greater network.

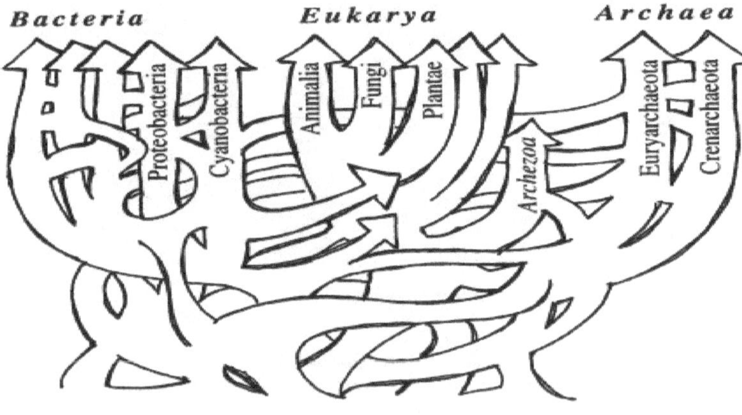

The tree diagrams of today, even if less aesthetically pleasing than Haekal's efforts, suffer less from these faults. They are increasingly using time as the x rather than the y axis to give the layout we associate with a flow diagram rather than a "tree" An alternative approach which would seem to have some advantage would be to plot time radially from the "common ancestor". The problem with this approach is that the diagram rapidly becomes too densely populated for easy interpretation. This difficulty has been recently overcome in a wonderfully elegant manner by James Rosindell of University College, London. Drawing upon the capability of interactive computer displays, a fractal inspired conic helix provides a remarkably intuitive way of navigating through a vast amount of evolutionary data to whatever level of detail may be required. It is called "One Zoom Tree" and is to be found online at www.onezoom.org. It is also a thing of beauty and not to be missed. As evolutionary "tree" diagrams go this is probably the ultimate approach and bears great promise in other fields as well. It also has the advantage of a "lateral flow" presentation in which time is essential the third dimension and does not lead to the basically emotional problems that Gould and others have with humans (or any other species) being at the "top of the tree".

As pointed out in a previous chapter, the tree diagram, even with all the merits of One Zoom Tree, cannot properly depict the evolution of the intricate network which is the biosphere. There are various measures of biological intricacy at any point in time. It can be described in terms of overall biomass, living biomass, or in terms of the intricacy of individual organisms. Or even as the biospheric network integrated over time. In respect of instantaneous total biomass we can only make meaningful estimates at the extremes of time. At the beginning of biology, let's say the time of the first free living cells, we can safely say that the total population was very low. One kilogram would seem to be a wildly generous estimate. One microgram perhaps more realistic. And, although there are, without doubt, many creatures, particularly microbes, yet to be discovered, present estimates of total biomass are already well in excess of 1000 billion tonnes of the carbon content alone. Although we can expect an overall exponential curve to apply we know that biological evolution, although essentially incremental, does not follow a steady course, proceeding quite slowly at some times and very rapidly at others. We see this even in the various expansions and diversifications of humankind. An overall exponential growth, but there have also been long periods of relative stagnation followed by spurts enabled by one or more aspects of the biosphere reaching a critical level. Rather like a river slowly but surely filling a natural dam until it bursts. Although the growth of the network is also perturbed from time to time by catastrophic events, its resilience is such that these temporary set-backs seem to open new compensatory niches

Another measure of biological intricacy is that of diversity, usually expressed in terms of the prevailing number of species. Going back to the earliest generation of free-living cells, the presently most widely accepted guess is one. Even those of us who would be reluctant to accept the single potential ancestor model would be hard pressed to propose more than, say, 1000 contenders. As before, in comparisons like this, a few orders of magnitude are of little consequence. We can be pretty sure, however, that the number of "species" would be very tiny compared with the 2 million already known and the 10 million estimated today. (Not to mention the huge number that have become extinct)

Yet another dimension of intricacy concerns that of the individual organism. This is not always an easy matter. We can certainly say that a dog is far more intricate than an amoeba. And that a shark is more intricate than a daisy. But is a cow more, or less, intricate than a horse? We have to tread very warily with comparisons of this kind. There are many traps to fall into. One such, very common among popular science writers is the blithe assertion that "The human brain is the most complex entity known." Which is nonsense! The most complex entity known is the entire universe. Or, more accurately, that part which is knowable. A good candidate for the second most complex entity might be the sub-set thereof , a galaxy. Of which a solar system is another subset. Then possibly the subset of that which is the biospheric network and includes biology. Only at this level of complexity do we find the human brain.

And even here, have we any good grounds to believe that the human brain is more complex than that of the whale or the elephant or the electric fish? It is a field which seems not to have been sufficiently researched. And how do we quantify neural complexity anyway? Certainly not by mass. The number of brain cells will come into the equation, but is not sufficient in itself. The size, internal structure and variety of neurons will be factors, And, very importantly, the extent and manner of interlinking, perhaps reflected in the number of synapses? As well as the many non-neural aspects, the plethora of neurotransmitters and hormones required for the brain to function.

It might be (and often is) argued that "obviously" our brain is more complex than all others because we do heaps of stuff that other animals clearly don't. That all other animals are inferior to us. This attitude, whether tacit or overt, is very wrong. Humans are ***adapted*** to do human stuff. Their neural configurations being modified accordingly. They have outstandingly large memory capacity which, together with the very extensive innervation of communication channel organs, enables our imaginations and the ability to share this imagination by means of language. It happens that this particular propensity results in the prolific production of artefacts, some of which enable us to occupy new niches. Dog's brains are adapted to do dog stuff. Their remarkable olfactory sensitivity and selectivity far outstrip ours. This is unlikely to be solely a matter of better receptors. Considerable processing could well be involved. Are dog's brains more complex than ours? Probably not! But that of the whale, with the extremely demanding task of echolocation, might well be. Or that of the electric fish, which has a remarkably different geometry from ours. Or that of the elephant, among whose computational tasks are the myriad complex interactions with its trunk, an organ whose wide range of capabilities are usually greatly underestimated, as well as its particularly high olfactory, auditory and vocal capabilities. Or even the octopus, with its more widely distributed nervous system that enables the coordination of its eight limbs as well as the trickiness of cleverly varying its chromatophores so as to closely match its background.

There is no sound evidential basis for the assumption that our brains are uniquely complex, even among those of other vertebrates. All species have their special abilities. As pointed out elsewhere in this book, our particular set of unique features are:

- Unusually complex sensorimotor innervation of the hands.
- Unusually complex sensorimotor innervation of the lips, tongue and larynx areas.

As clearly depicted here:

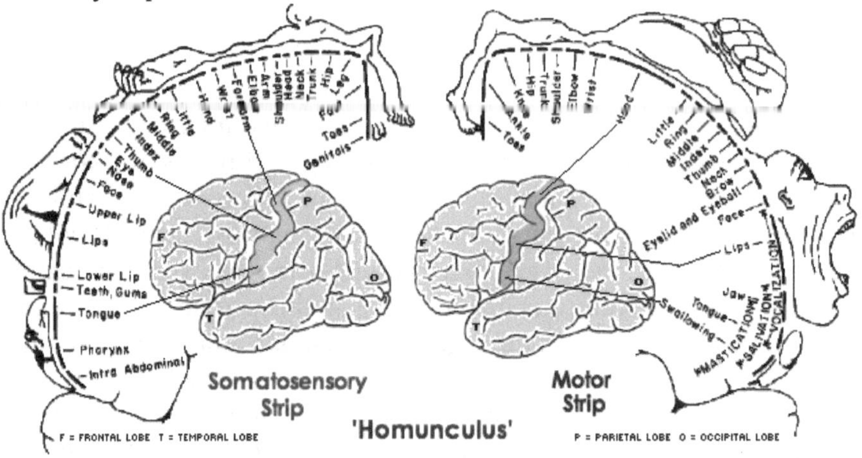


This diagram is an experimentally derived representation of the relative levels of sensory and motor innervation serving the various parts of the body. These levels being depicted by size.

This inordinately high degree of innervation of the hands and the vocal apparatus are the particular attributes which underlie the unique set of behaviours exhibited by the snoutless ape. The intensive control and feedback wiring to the hands being enabling and, in turn, being further enhanced by the use of tools. A capability inextricably entwined with the transmission of imagination as graphic language, which, as we will come to see later, is actually the most significant communication channel for the sharing of imagination.

The homunculus also highlights our other main communication

channel, the vocal apparatus, which is better suited for real-time and early acquisition applications of imagination transfer. The alert reader may have spotted that the hands, although having considerable input sensitivity and selectivity, are most often used as transmitters of language and the that the vocal apparatus is used ***only*** for transmission. Reception is effected mostly by the eyes and ears, respectively. So, although the eyes and ears are indeed enlarged somewhat on the homunculus this might seem less than expected. This can be explained by the realisation that the somato-sensory and somato-motor strips are essentially switching stations that access remote locations. The ear and the eye, on the other hand are directly integrated into the neocortex. Indeed, the memory intensive processing required for decoding, reaction to or reconversion to local imagination of these complex visual and aural inputs must be a major function of the human neocortex and largely responsible for its larger size. For comparison we will bring on stage again our old friend, the star-nosed mole, with his own particular highly specialised talents.

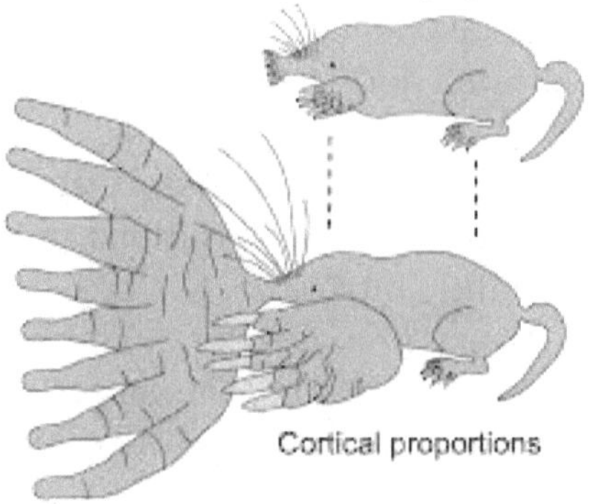


**Molunculus**

# Eukaryosis

One of the most intriguing issues is the apparently rather sudden emergence of an entirely new level of biological intricacy from a network that seems to have been previously the exclusive domain of the prokaryotic bacteria and archea. The Eukaryote engine is a radical departure inasmuch as that kind of cell, as we typically see it today, has an entirely distinct set of features. The most conspicuous is a clearly defined nucleus in which the protein specification molecules of DNA are shielded within a membrane. Access to which is normally limited to RNA and other molecules having security clearance. Among the many other special devices used by organisms of this set, important components are the intracellular energy transducers. In animals, the mitochondria. In plants, chloroplasts. It is almost certainly the introduction of these dedicated organelles which enabled the eukaryotes to evolve to take advantage of the opportunities provided by larger size and dedicated multicellularity.

The archea and the bacteria, as discussed above, had, over time, adapted to a vast range of habitats. Variations in their chemistry and geometry increasing the intricacy of the biospheric network enormously in both diversity and extent. There was, however, one frontier left unexplored, that of size. The largest known bacteria are about half a millimetre long. And they are truly giants among their kind.

The problem with powering larger organisms is one of logistics. In the case of animals, providing a supply of electron-rich nutrients to all the cells. Either by direct diffusion or by a reticulated circulatory system. The second problem, that of getting oxygen in far enough to complete the circuit by accepting low energy electrons. It is similar to the problem of building an elephant-sized ant. The ant gets its oxygen by diffusion through trachioles in its "shell". Because volume increases as the cube of size, the scaling up of an ant soon runs out of air and new systems have to be put in place. Like the lungs and other clever bits and pieces the elephant uses to get oxygen spread around its bulk.

So, rather than simply providing a supply of electrons by means of proton pumps embedded all over its surface, in the eukaryote nature has found a way for this to be internalized by transferring the transduction function to the organelles. Thus opening the door to much larger cell size and true multicellularity with differentiation and Integration of the swarm. A great leap forward in entirely new opportunities, new niches into which the network of chemical and geometric intricacy would expand.

The million dollar question, of course, is how did this come about? It is now pretty well agreed that the ancestor's transducers, the organelles, were originally free living prokaryotes who somehow managed to find themselves inside another organism. One proposal is that one organism (probably an archean) swallowed another (probably a bacterium) for lunch. Having somehow evaded digestion, the swallowee making itself at home. There seems to be a problem with this because, as far as we know, no prokaryote seems to have evolved this kind of mechanism for having a snack. Quite possibly because of the same energy distribution reasons that limit maximum size. The cell membrane being a very busy place, kept flat out with respiration and other requirements. Functions which in the eukaryote are largely subcontracted to mitochondria and other organelles. Another scenario proposes that the intruder originally broke in as a stealthy marauding parasite, only to subsequently mend its ways, having found it a better deal to cooperate with its host.

I have been waxing metaphoric, of course. Whatever the evolutionary details, the deal was not struck overnight but mostly occurred over countless generations, during which time the organelle became increasingly more dedicated to electron supply, casting off most of the now redundant clutter of DNA not relevant to its new duties. Some of these genes picked up by the host to use for new structures and others chucked out. While there is contention over details, this overall picture is widely accepted and makes good evolutionary sense.

There is one strange observational feature, however, which is a puzzlement. For, as far as we know, during the long transition from "parasite" to fully paid up member of the cellular association, there was no spin-off. No intermediate organisms that would go their own way on parallel paths. At least, with one possible recently claimed exception, there is a dearth of evidence. There is always the possibility, of course, that such breakaway movements did occur but couldn't cut the mustard and became extinct. Nick Lane, who is particularly articulate and sharp-witted in these matters, points out that the archezoa, formerly thought to be protists, turned out to be single celled eukaryotes. They successfully occupy niches that would have been tenable by early protist metazoans on a parallel path. That makes sense up to a point, but it seems to me that because such niches exist now does not necessarily imply that they existed way back then, two billion years ago, when the network was far less intricate. Another distinct possibility is that these less able descendants of eukaryotic experiments could have later have actually been displaced by those very protists who occupy the niche today.

In competition with the engulfment models, either pro-active or parasitical, is the "hydrogen hypothesis" an ingenious outlier which proposes a much earlier derivation of the mitochondrion from hydrogenosomes. These are organelles observed in some modern anaerobic eukaryotes which, in a symbiotic coupling with methanogens, derive electrons in the course of an overall conversion of pyruvate to methane. They are also closely associated with archea. This model involves an archeal transformation occurring ***before*** significant oxygen availability and thus at a much earlier time. Well back in the era of the "boring billion", the years in which biospheric evolution seems to have proceeded at a very slow pace.

While, much of the time, the changes to biological organisms seem to take place by tiny steps, we now have evidence that, even at the genome level, "macro-mutations" sometimes occur. Chromosome doubling and even genome doubling are observed phenomena. In such cases the organism's "risk" of niche exploration is reduced and the rate of overall reproductively successful diversification can be increased accordingly. The "spare" chromosome in this way being able to act as an experimental test-bed, while the original continues to perform its well established everyday functions. The rather sudden emergence of the flowering plants may be at least partly attributable to such an event. Cell fusion is yet another example of more drastic change.

Even these cannot compare to the prodigious macro-mutation event leading to what has been called a "hopeful monster" that has been advanced by some to solve the puzzle of the seemingly unique and sudden advent of eukaryosis. We should certainly be reluctant to take this kind of scenario too seriously. Over a very wide range of networked system we find that the probability of events are best described mathematically in terms of some kind of power law. Major catastrophes being very rare and small events of all kinds being very common. Biological evolution appears to occur, for the most part, in "the long tail". Extraordinary claims require extraordinary evidence, and, for an "all bells and whistles" unique emergence of eukaryosis, that we don't have. Even if it were the case, however, it would not invalidate the underlying incremental theme pursued in this book. It might just conceivably have been "an accident waiting to happen". But it lacks the quality of consilience. It does not quite gel with the "big picture" and must thus remain suspect.

In a previous chapter we noted the importance of cooperation to the biospheric network. Somewhere around two billion years ago this most remarkable and still very mysterious example of cooperation occurred. From an engineering perspective one can quite easily appreciate the efficiencies and potential for the newer high levels of diversification provided by the inclusion of task-dedicated organelles within the cell. Specifically the mitochondria. Which, together with the advent of a sufficient level of oxygen, made energetically possible the previously untapped scaling up of organisms far above the size constraints inherent in the prokaryote modalities. The manner in which it achieved the required level of sophistication for its viability is, as we have seen, perplexing. For, according to our best current understanding, there appears to be little room for evolutionary intermediates. Rather than the baby steps which can be attributed to most other examples of bio-evolution, this appears to be more of a giant leap. Which sets off an alarm that perhaps something, somewhere, is wrong. There is one recent discovery which, if confirmed, may lead to new ways of thinking about this. New micro-organisms are being discovered all the time. Some in the most unexpected habitats. Deeply embedded in rocks, in polar ice caps, in the high temperatures and pressures of sub-oceanic thermal vents. Once biology gets under way, the network seems to extend its adaptive tendrils to explore all possibilities. What appears to be an intermediate organism, Parakaryon myojinensis, has recently discovered by a team led by Masashi Yamaguchi of Chiba University. Electron microscopy indicates that this has the nucleus characteristic of eukaryotes but it is bounded by a single rather than double nuclear membrane and contains endosymbionts that resemble bacteria not yet differentiated into mitochondria. This would strongly favour the model of a gradualistic evolutionary pathway for the emergence of the eukaryote. This organism was found rather serendipitously in a study of a type of sand worm. Which, quite intriguingly, is found in the vicinity of our old friends, the hydrothermal vents.

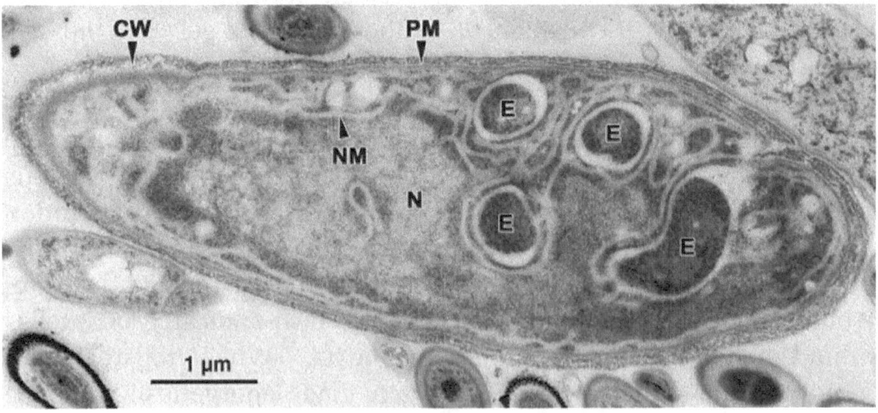


*Parakaryon myojinensis*

This find, of course, casts very serious doubt upon the "hydrogen hypothesis.

In general, speculations on the origin of eukaryotes are many and varied. For instance, Martin Brasier of Oxford University remarks:

> *"I argue that the boring billion was the anvil on which the eukaryote cell was forged. If so, then modern eukaryote cells could be the product of geologically rare conditions."*

# Cooperation

A characteristic of biological evolution is to learn by its mistakes. Even at the level of the most primitive functioning evolvable cell we see that biology is characterized by a cooperative society of macro-molecular assemblages. Those of the cell membrane that provide input and output facilities, of the many enzymes that read and manipulate the data source which stores the recipes for proteins, the protein assembly mechanisms and those that install these proteins all being important examples. A pattern of cooperation that is to be found at all levels of the biological realm.

Ironically, we can reasonably include predation and parasitism as special modes of cooperation in which one of the organism is the "unwilling" and unwitting partner that assists with the process of the furtherance of molecular intricacy. No organism can properly be considered in isolation from the biosphere upon which it is utterly dependent. For bio-evolution is not that of a species but of a highly integrated and interdependent network of organisms. A world wide bio-web. In the middle of the last century a project inspired by a speculation of Schrödinger culminated in the work of Franklin, Gosling and Wilkins which firmly established the structure of DNA as well as the means by which this oligomer stores genetic information. The tsunami of enthusiasm for matters genetic which followed this discovery inspired attempts to explain all evolutionary phenomena in terms of the gene. Although some quite important models were to emerge from the gene-centred approach, this rather naive oversimplification of evolutionary processes led to far too much emphasis on competition as the driver

While competition is, without doubt, a significant component of biological evolution, the very important other side of the coin, cooperation, has in more recent times been brought to the fore by such workers as Lyn Margulis and, more recently, James Shapiro. For bio-evolution is not that of a species but rather that of a highly integrated and closely interdependent network of many organisms. Even the nucleic acids can be viewed as cooperatives of nucleotides which work in concert with the proteinacous enzymes in specification of a phenotype, rather than a simple one to one mapping of characteristics.

After all, the gene is a mere section of code, a chunk of data, the recipe for a protein, and, although an essential link in the chain of chemistry that implements replication, it is in no way to be considered the primary driver or even the most important component of bio-evolution. For, without the compatibility of interaction with the bio-web, within which it is, of necessity, expressed, of networks which are essential for it to be properly represented, the gene is impotent. As pointed out in chapter six, the concept expressed by the mantra "DNA makes RNA makes protein" is hopelessly skewed. A more realistic re-phrasing of this being: "Proteins (enzymes) act upon DNA as a template to make RNA. As well, in accordance with intracellular and extracellular requirements, they then use RNA to make more protein. They also make further copies of DNA. The enzymatic and structural equipment required for these processes being directly propagated down the cell line". In the case of the multicellular eukaryote, any particular cell line eventually becomes senescent, but the process then begins anew using the intracellular inheritance provided, initially, by stem cells.

Selection is not a one way street, nor is it executed by any single mechanism. No more than it is effected by the global bio-web alone, but also by interaction with the prevailing non-biological chemistry and physics of the Earth's surface, the oceans and the atmosphere. These, in turn being dynamically modified by the bio-web itself. The constant interplay of all these multitudinous interactions ineluctably exploiting the underlying structure of our universe to ratchet forward ever-increasing chemical and geometric intricacy and diversity.

Another well established myth associated with the nucleic acid protein recipe collection is that associated with "selfishness". The naïve model from which this stems certainly has some merit. But, again, like the Bohr motel of the atom its utility is limited and should not be given too much weight. After all, the gene is passive. It is expressed solely by the cell machinery in the context of other genes and the expression can be modulated by signals that may emanate from many levels of the biospheric network. To take a simple example, eye colour is not expressed in the liver or the heart. The genome is subjected to extensive translation by the cell's machinery according to its location and status within the network. While competition certainly plays a part in locating an evolutionary niche, cooperation between organisms is at least as important. Indeed, the ubiquity of symbiosis. including the particularly intimate integration of organisms that has given rise to the eukaryote, suggests it is better considered to be the main player. James Shapiro is a strong advocate of this more sophisticated view of biology, His work demonstrates instances of elaborate co-operative behaviour in bacteria including self-sacrifice for the benefit of the group.

Even at the level of RNA, theoretical studies suggest that cooperative RNA replicative networks compete successfully against simpler "selfish" autocatalytic cycles. Providing a positive selection pressure for increased intricacy. A seemingly scale invariant gross effect which can be detected at all levels of biology and, indeed, in sociology and technology. In encounters between human societies, those more complex have generally prevailed over the more primitive. In technology, be it weaponry, textile manufacture, agriculture or transport, simpler devices usually give way to more intricate, more highly organised systems.

# Contingency versus necessity

Stephen Jay Gould is noted for his passionately promoted conviction that the emergence of new species is primarily accidental. This viewpoint was largely inspired by his encounter with the immensely rich fossil beds of the Burgess shale in Canada, where have been found the enormous diversity of organisms that arose during the "Cambrian explosion". The vast majority of which have long become extinct and leave no trace in present organisms. We have already looked at the phenomenon of completedepascency, which can be expected to occur in any system subject to the thinning-down effect of natural selection.

Gould's rather odd interpretation of this, however, was essentially that, of all these creatures, some having very bizarre and complex morphologies, each of these had an equal chance of surviving to this day. In other words that their fortunes were in no way favoured or inhibited by the state of the network in which they found themselves at any point in time. Implying that the network itself had no directionality. Which is contradicted by the fact that it very observably has. As with many before and since, he had fallen into the very common anthropocentric trap of equating directionality with intentionality, which is actually a whole 'nuvver kettle of fish. In point of fact, his contentions were not really quite as naive as the well publicised out-take with regard to the "tape of life" thought experiment suggests. Unlike the many followers who were overly influenced by this surmise, he seems to have been actually well aware of the rather obvious fact that morphology is by no means solely dependent on contingency. In chapter four of "Wonderful Life" he writes:

> *"Am I really arguing that nothing about life's history could be predicted, or might follow directly from general laws of*

> *nature? Of course not! The question that we face is one of scale, or level of focus. Life exhibits a structure obedient to physical principles . We do not live amidst a chaos of historical circumstances unaffected by anything accessible to the "scientific method" as traditionally conceived."*

The question we have is indeed one of scale and focus. But it is also a question of where we draw the line. Strangely, Gould blithely accepts that

> *"the origin of life on Earth was virtually inevitable, given the chemical composition of early oceans and atmospheres".*

He inconsistently assumes no contingency here, even though the chemical composition of early oceans and atmospheres is far from certain. In the account being developed in this book we also reach the conclusion that abiogenesis was pretty well inevitable, but for a very different reason. That of the evidence provided by strong evolutionary patterns which are observed in our part of the universe., rather than the lack of evidence regarding early atmospheric and oceanic conditions. Or, for that matter, the not yet fully understood "physical principles of self-organized systems"

Gould's many literary efforts are informative and interesting. but his various interpretations tend to be distorted by parochialism and an obsession with establishing that the emergence of a species such as ours is a mere fluke. Sometimes he very deftly catches the ball but fails to run with it. As with his analogy of "spandrels". This is an architectural term for the gaps which are, of necessity, present in a structure when, for instance, a dome is mounted on top of a circle of arches. They perform no intrinsic function in the building and their provision is not of choosing. Rather, they are thrust upon the architect, who often opts to give them a secondary function, that of containing ornamentation.

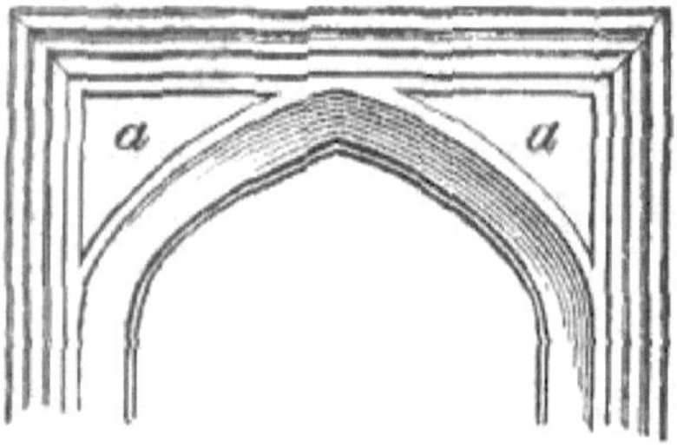


*a, a,* Spandrels.

But what Gould failed to do was to follow the very good analogy through properly. To convert the analogy taken from a static system to one that is dynamically evolving. Which then corresponds exactly to the category of pathway for which I use the term "spin-off".

Unfortunately, the nature of the universe in which we find ourselves is not to be discovered by simply reflecting upon the fossils of the Burgess Shale, fascinating as they may be. His famous "replaying the tape of life" eidolon has already been rebuffed in one way already in chapter nine, but the many well-established examples of convergent evolution give even further cause for rejection of Gould's flawed contingency contention.

It is claimed that more than forty versions of the complex lensed eye have evolved from the basic photoreceptor quite independently.

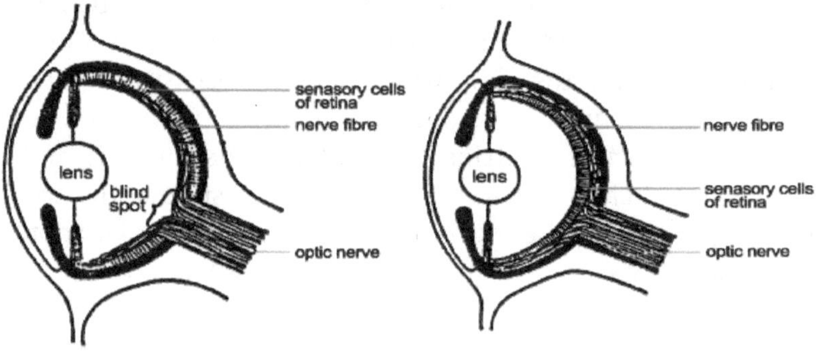


*Derived from art by Jerry Crimson Mann*

The shape. of an aeroplane is constrained by aerodynamics. As are the materials used for its construction. Mercury, sulphur, nitroglycerine and candy-floss being quite unsuitable. Similarly, the essential geometries and construction materials of other devices are also constrained by function.

*Saxophones are not suitable for powered flight, nor motorcycles for dress-making. The chemistries and geometries involved in their construction differ.*

The well-established phenomenon of convergence, whereby similar morphologies (geometries and chemistries) are found to have evolved from very distinct lineages, is widespread, and also exhaustively documented. It is yet another nail in the coffin of the ideology of a random, directionless process. As above, the number of engineering solutions available to any functional requirement are constrained by such variables as temperature, viscosity, availability of fuels and oxidants, mass, light intensity and frequency and many other parameters of the network within which bounds the organism operates. It is further constrained to various extents by the materials and information for their reconfiguration that have been handed down from its ancestors. So it is not at all surprising that, where an ecological gap occurs, it is often filled by entities having a geometrical or chemical similarity to those to be found in an equivalent ecological gap elsewhere. Where regions become isolated for a long period, various aspects of the prevailing ecology will, over time, become reconfigured along corresponding lines. The classic example is the native fauna of Australia, where the marsupial species, free of competition from the slightly more efficient mammal, have strikingly similar equivalents to major mammalian species.

It is true, of course, that the forces that shape convergence of traits do not always come up with precisely the same result. This is simply because the organisms which are converging evolve from different spatio-temporal parts of the biospheric network. Their structural starting-points are different. The available materials (information) with which any organism starts the process necessarily limits the ways in which selection can act so as to implement the convergence. Sometimes the initial "set of parts" which are available for variation or exaption is insufficient for the process to get anywhere near completion. Which gives us the phenomenon of the "adaptive peak".

# "Bad Design"

In his popular book "Evolution as Fact and Theory", Stephen Jay Gould remarks:

> *"Evolution lies exposed in the imperfections that record a history of descent. Why should a rat run, a bat fly, a porpoise swim, and I type this essay with structures built of the same bones unless we all inherited them from a common ancestor?"*

So far, so good. But he then continues with the bizarre statement:

> *"An engineer, starting from scratch, could design better limbs in each case."*

Which, of course, is utter nonsense! No engineer or team of engineers, even having the very extensive range of resources, ideas, technologies and materials available to us today has the capability to produce a prosthesis having anything like the very sophisticated functionality of the human arm and hand. Nor, for that matter, that of the locomotory mechanisms of the rat, bat or porpoise. Let alone "starting from scratch". Bearing in mind Carl Sagan's perceptive remark to the effect that

> *"If you wish to make an apple pie from scratch, you must first invent the universe."*

Sadly, Gould is not alone in this silliness. The literature is liberally sprinkled with similar comments. Among the many examples are:

Jerry Coyne, who points to the circuitous route taken by the recurrent lingual nerve from the brain to the larynx as

> *"one of nature's worst designs,"*

Richard Dawkins, with regard to the same feature which, in the case of the giraffe, involves a deviation of around 5 metres:

> *"Obviously a ridiculous detour! No engineer would ever make a mistake like that."*

So are we then to believe that we have among us some human engineers who would have the capability of "designing" a giraffe? Even a slightly "imperfect" one? Or that our human engineers (presumably "intelligent") never make mistakes? As in the case of the seals of the Challenger space shuttle, perhaps? Or, even more pertinently, that our human engineers never make compromises? Compromise, as emphasized by Henry Petroski in his book "Small Things Considered: Why There Is No Perfect Design", is in fact very much the norm in the "design" efforts of we humans. Not only are arguments such as these specious, but they also tend to engender an unwarranted acceptance of extreme human exceptionalism. The human engineer, far from being able to design a better giraffe, is not even capable of producing a single cell! While it is very proper to refute the entirely unevidenced claims of creationism and of "intelligent design", these irrational sentiments clearly do not cut the mustard.

Another feature often subjected to derision of this kind is the "poorly designed" human arrangement that diverts food and water to the oesophagus rather than the trachea. A "design flaw" which it turns out is crucial to the primary niche involving endurance hunting occupied by our not very distant ancestors as described in chapter twenty-three. The tacit anthropocentric mind-sets revealed by comments such as these from many professed evolutionists are indications of just how difficult it is for us to discard the notion that we are separate from nature and that our minds are somehow "supernatural". An almost universal human conceit which, despite all the evidence to the contrary, we seem loth to forgo.

# 22

# AD INFINITUM

*Inspired by a verse penned by Jonathan Swift, the mathematician Augustus De Morgan whimsically proposed that:*

> *"Great fleas have little fleas upon their backs to bite 'em,*
> *And little fleas have lesser fleas, and so ad infinitum*
> *And the great fleas themselves, in turn, have greater fleas to go on, While these again have greater still, and greater still, and so on."*

*This was, of course, intended as a bit of fun but it also serves to illustrate the inherent limitations in the use of mathematics to describe the real world. In which the big picture of science indicates very strongly that the size of fleas is seriously constrained by physical and chemical parameters. Elephant-sized fleas are only to be found in the less thoughtful regions of fiction. The same considerations apply to nanofleas.*

*Less obviously, infinities, a wonderful vehicle for travels in the realms of pure mathematics have questionable value for the scientist in attempts to describe external reality. Many concepts used in the very simple language of mathematics do not correspond to observable phenomena.*

*If treated as hypotheses they fail on all counts. A mathematical point is dimensionless. As such there is no way that it can be observed or detected. Similarly with the one dimensional straight line. Of which, we find that even the* two dimensional approximation is, in the real world, never truly straight at small dimensions.

Again, we have never observed any two dimensional plane object. None of these things actually occur in nature. They are not part of science. They are mere artifices, idealizations which, properly applied, are enormously useful for helping to interpret relatively simple systems and phenomena. But scientific endeavour starts and ends in the real world. The language of mathematics simply oils the wheels.

Albert Einstein's observation that:

> *" As far as the laws of mathematics refer to reality, they are not certain; and as far as they are certain, they do not refer to reality."*

conveys a similar sentiment. In chapter seventeen, with reference to Gould's "tape of life" thought experiment I raised the question:

> *"Among the billions of artefacts they produced, would they include an Eiffel Tower? Would they have a replica of the Vatican, complete with a reproduction of the Sistine Chapel and its ornate ceiling? Van Gogh's Starry Night? Big Ben? Beethoven' Fifth Symphony? "*

And answered in this way:

> *"We would have to consider the recurrence of even one of these, having a vanishingly small probability."*

Now there are those that would claim that, since our universe is infinite, the recurrence of each of these (and everything else) is certain. All things must have exact duplicates (and, for that matter, triplicates, quadruplets...)

*The assumption of real world infinities is fraught with irresolvable problems. Louis Pasteur had strong feelings on this matter:*

> *"I see everywhere in the world, the inevitable expression of the concept of infinity.... The idea of God is nothing more than one form of the idea of infinity. So long as the mystery of the infinite weighs on the human mind, so long will temples be raised to the cult of the infinite, whether it be called Bramah, Allah, Jehovah, or Jesus.... The Greeks understood the mysterious power of the hidden side of things. They bequeathed to us one of the most beautiful words in our language—the word 'enthusiasm'—En Theos —"A God Within." The grandeur of human actions is measured by the inspiration from which they spring. Happy is he who hears a god within, and who obeys it. The ideals of art, of science, are lighted by reflection from the infinite."*

*Be that as it may, and disregarding the religious overtones inevitable in his day, a universe without infinities is unimaginable. The trouble is that the same can be said for a universe with infinities.*

*The relationship of the language of mathematics to the real world is considered in appendix B*

# 23

# HUMAN EVOLUTION

Somewhere in one of Douglas Adams' books I seem to remember Arthur Dent remarking "That's why nothing ever seems to make sense" And in the title of a 1973 paper, Theodosius Dobzhansky famously observed that

> *"Nothing in Biology Makes Sense Except in the Light of Evolution"*

This is indubitably the case with our particular peculiarities. The multiplicity of bizarre, perverted, pointless, iniquitous and seemingly counter-productive aspects of human behaver are a puzzlement. But the enigma can be resolved by taking it outside our usual human frame of reference and instead interpreting more objectively in terms of nature's evolutionary machinery. Whereupon most of our strange peccadilloes become entirely explicable.

Our quest to interpret the patterns found in nature is often hampered by inadequacy of data. Perhaps nowhere is this more so than in the field of palaeontology. Only a very small proportion of biological remnants become preserved by the various processes of fossilization and, of those that we have been fortunate enough to have found, many are mere fragments that allow of various interpretations. Methods for determining ages of fossils fairly accurately have been developed but, because of the continual branching of evolutionary pathways, there is, of course, no guarantee that any particular example can be considered representative of the clade of which it is a member. Thus, while the fossil record can give some useful clues as to the developmental history of a creature, any assumptions regarding any particular path must be considered to be tentative.

Fortunately, we can call upon other disciplines to help connect the dots. Biochemical studies, particularly the relatedness of DNA found in the cell nucleus. Also of the much simpler and more rapidly changing DNA found in mitochondria (passed almost exclusively down the female line). A very useful, if somewhat controversial tool. Also, quite importantly, we can call upon the general background of well established bodies of knowledge such as physics and chemistry, the "big picture", to analyse function. In reconstructing that path which has led to the human machine we can use the principles of engineering, along with quite reasonable assumptions based on the overall geological and biological conditions pertaining over the last several million years .

It is sometimes easier to solve puzzles "backward". Adopting this approach we can start by considering the salient anatomical and physiological features of the modern human. From these we can glean quite strong clues as to the ecological niche(s) within which adaptations have taken place. We can usefully look at the specifications of this machine to flesh our the evolutionary trajectory in terms of the niche(s) in which it has adapted to function. This is a valid and quite powerful heuristic approach. For instance, the specification does not include wings or vestiges thereof. It is quite certainly not an avian.

**Bipedalism:**

The simplest layout for a terrestrial vertebrate in terms of stability seems to be that of a leg in each corner. In fact that basic body plan appears to be a legacy from the age when fishes first dragged themselves out of the water. While unused niches exist, or some new ones arise, the processes of biological evolution seem to unerringly find new strategies to exploit them. Sometimes the quadrupedal model was to become exapted in relatively minor ways such as, in the case of the cat family and the squirrels, the use of claws to climb trees. All mammals exhibit difference in form and function for their fore and hind limbs. In the case of many monkeys and apes, specialisations of fore and hind feet are such as to allow grasping of branches.

In some cases, further simplification of the body plan occurred, as in the case of the snakes and exclusively aquatic mammals, with the investment in legs no longer required. (It is believed that the recipes required for these various appendages are not usually deleted from the genome but rather modulated or inactivated and hoarded for a while.) Sometimes being pressed into use again if circumstances demand. Various other adaptations to fill niches, to meet prevailing conditions, involved the freeing up of the fore-paws for functions other than locomotion. Effected by standing upon the two back feet. Giving an extended vertical foraging range and/or surveillance for predators. Just as commonly employed by various species today. The Meerkat, Prairie Dog, Bears and the Kangaroo are examples. In fact, it may well be the case that obligate quadrupedal terrestrial vertebrates are in the minority.

Bipedalism, even the locomotory version, was prevalent among the dinosaurs and is universal among the birds. So it should thus come as no great surprise that of the many and varied successors that sprang from the small rat-like therapsids, and among those classed as primates, some adapted to various opportunities provided by life in the higher levels of the forest while others filled the fairly new niche provided by the bushes and fruit trees characteristic of the rosaceae. For which not a tree-dweller but, rather, a predominantly terrestrial bipedal machine with a modest tree climbing ability would seem to fill the bill. (The proposition that our lineage may have been habitually bipedal since the time of the therapsids has been given very little consideration so far. This appears to be a result of naive acceptance of the "when we came down from the trees" myth that is reinforced by confirmation bias rather than hard evidence. However, rather than interrupting the flow of narrative here, however, I will take up that particular contentious cudgel in appendix A.)

The present human, the chimpanzee and the gorilla, all having specialized in some respects, nevertheless more closely meet this primitive ancestral specification than most other species. So here we have a very plausible and more straight-forward explanation for the origin of bipedalism in our particular lineage. An optimal means of using of a vertical section of available nutrients. Subsequently, the descendants of this bipedal frugiverous and insectivorous forager would go their own ways, the chimpanzee to a slightly higher stratum of the forest. The hominid line adapting to the newly presented opportunities of the open plains and woodlands, while for the gorillas it was, to a large extent, business as usual.

**Locomotion Modes:**

Bipedal locomotion in the present non-human apes is very clumsy. And since this was probably also true of our early progenitors, considerable adaptations that included rotation and remodelling of the pelvis and of the knees and feet were required for the animal to be able to efficiently make its way in the world as an omnivorous wanderer of the woodlands, beaches and plains. Indeed, the family of australopithecines, although being essentially herbivorous browsers who still retained the snout, had already some of these more efficient locomotory features.

Some quadrupeds have evolved two or more distinct modes for locomotion. The horse, for example, uses the walk, the trot, the canter and the gallop. Adaptations of this kind reflect an ecological niche such that selection favours the ability to travel at a range of different speeds with optimal efficiency. These gaits performing a function somewhat akin to that of a gearbox in a motor vehicle. This can be important to both predators and prey.

The facility is far less common in terrestrial bipeds. But the human machine, although having only three basic gaits, the walk, the jog and the run, has also a host of other gaits such as the hop, skip and jump. Which also makes some of us such excellent dancers. Some of these gaits are most likely of more recent origin and reflect the greater levels of coordination arising from extensions of our central nervous system. The locomotory flexibility of this hominid machine, which, by virtue of its particular configuration of pelvis and knee, is also able to very usefully crawl, is suggestive of a varied and demanding early developmental niche.

**Thermoregulation**:

Another feature of the human machine is the unusually efficient set of heat dissipation arrangements. Among the various components of a sophisticated evaporative cooling system, we find the eccrine sweat glands. An ancient system which is restricted to land based primates, a further indication of our terrestrial, rather than arboreal, heritage. Also, we find far less body hair than most mammals. A further aid to evaporative cooling. Particularly for a predominately vertical biped of small tubular cross-section. Vertically reducing the "ground-effect" whereby wind-speed falls off rapidly with height. The quadrupedal camel uses its long legs to similar advantage.

The notable exception to the nakedness of the human ape being the head. Where protection from the direct overhead radiative heat of the sun is required. Cooling of the CNS is aided by additional specialised arrangements for blood vessels serving the brain. (Much as you may have a dedicated fan to cool the CPU of your computer). Another feature, shared with the gazelle and the camel, is the ability for the flow of venous blood that has been evaporatively cooled to be switched into reverse and sent to the brain. While, for these animals, the cooling occurs in special nasal cavities, the human uses sweat -cooled facial surfaces. The elephant, too has a specialised brain cooling system. Required to cope not only with the thermal challenge presented by its environment but also by the information processing requirements of its great body mass, as well as the demands of that most sophisticated organ, the trunk. In this case the two very large ears are the primary heat exchangers. In accordance with the best engineering practices they are vertically oriented (to reduce incident radiation), highly vascularised, far above the ground, of adjustable orientation and, when required, active flapping to increase conductive cooling.

We also find an extra layer of fat beneath our skin, this being thicker in the human female. Another gender related difference in regard to heat regulation is the much lower rate of sweating in the female. Yet again, it is found that the rate of sweating is increased by the secretions of the adrenal glands, Associated strongly with "fight or flight" or "chase and kill" situations.

So what is our engineer to make of this? In what prevailing conditions would this combination of properties be advantageous? One first clue comes from the correlation of evaporative cooling with adrenal secretions. This suggests that the additional cooling effected by sweating is associated with active rather than passive situations. "Fight or flight" or "chase and kill". A mechanism for the dispersal of internally generated heat during intense activity rather than when at rest.

The human has, in general, retained head hair to protect against solar radiation when incident overhead and at its most intense. Except in true desert regions, where reflection of heat from the sand surface is considerable. A strong indication that deserts were not regions for which these early hominids were adapted. It is not inconceivable that the adaptation of thick "frizzy" hair found on some Africans today could be a relic or replicate of that of early hominids. Sure, there are many other mammals of comparable size which have not found selective advantage in discarding hair in favour of evaporative cooling. Indeed, a good coat of fur, particularly if reflective, can provide insulation against radiative heat. The camel, for instance, is unable to use the strategy of nakedness to assist its heat-management because of the incident reflected heat of the desert sand. The very same reason present-day human desert-dwellers wear robes. The camel also needs hair on its back to protect the much greater surface area exposed to the overhead sun by virtue of its quadrupedal layout. Whereas head hair is sufficient to serve the same purpose for the bipedal human.

The human ape is not alone in its nakedness. The African elephant, for instance, faced with a very different engineering challenge, that of dissipating sufficient heat from its enormous mass, adopts a rather similar strategy for body cooling. Its wrinkled bare skin, very thick on its back, as well as greatly increasing surface area also uses evaporative cooling. Not by sweat, but by using water which is retained in the deep wrinkles after bathing.

Hair also gives protection from low temperatures. Which brings us to a problem with our hairless human. How does it cope with the cold? This problem, we find , is mitigated by the extra layer of fat which has been selected for. So we see that the loss of body hair and insulation instead provided by fat are compensatory with respect to ambient temperatures. What is different is that the cooling system has been transferred to the outside rather than the inside of the insulating layer. Enabling the prolonged high levels of activity in hot arid climates that are involved in endurance hunting, A method of predation in which presumably our ancestors excelled.

One small matter remains. How do we explain the thicker layer of fat and reduced sweat secretion in the female? This is entirely consistent with our present interpretation inasmuch as the rôle of hunter has naturally fallen to the male because males of almost all species are more expendable than females (When did you last hear of a male spider eating the female after mating?) Females almost always tend to be excluded from more hazardous and, in the present case, more vigorous activities. Their human reproductive success rates are better are served by extra insulation against cold than by a capability for dissipation of internally generated heat.

A fat layer also provides a back-up reserve of energy-dense triglycerides. Under normal conditions the main supply of energy in modern humans is glycogen (4.1 kcal/gm).1 which is derived from carbohydrates and stored in the liver. Endurance training of athletes, however results in an increase in our ability to oxidise the higher energy density triglycerides (9.3 kcal/gm.) This also makes available metabolic water, a mechanism important in that other heat endurance machine, the camel.

$$\textbf{GLYCOGEN } C_6H_{10}O5 + 6O_2 > 6CO_2 + 5\ H_2O$$
$$\textbf{FAT } C_{52}H_{104}O6 + 75O_2 > 52CO_2 + 52\ H_2O$$

There is a weakness in our present model. Today's marathon runner grabs a bottle of water to replace all that used in evaporative cooling from skin and lungs. When all of the eccrine sweat glands are working at maximum capacity, the rate of perspiration for a modern human being may exceed three litres per hour, and dangerous losses of both fluids and electrolytes can occur.

Stopping for a drink would not seem to be a practical proposition for an endurance hunter who must continuously pursue or closely track his prey. In the light of this it is perhaps reasonable to speculate that perhaps those predatory early hominids whose evolutionary trajectory we are trying to reconstruct had already acquired the use of the equivalent of a water-bottle. Maybe in the form of a gourd, stomach, bladder or even, as used by hunter-gatherers in the Kalahari today, ostrich egg-shells. Another possibility is that this species might have possessed more fat and more efficient biochemistry for its utilization. From the engineering perspective we have adopted, all these observed adaptations are strongly indicative of this human lineage optimized for the niche of endurance hunter. Indeed it is a lifestyle for which we are believed to have evolved two million years ago and that, even today, still survives in the bushmen of the Kalahari and the Tarahumara of Mexico. But we have not yet finished with adaptations attributable to this way of earning a living. The human ape is of a far slighter build than its contemporaneously evolved cousins, whose size and general body strength are greater. What is the evolutionary trade-off that allowed us to become so weak? Speed! It is easier to dump excess heat from a small body. It also requires less power for its propulsion, which is, of course, why we make racing bikes and cars as light as possible. The human who is in good nick enjoys the reputation of the world's second fastest runner over long distances in hot conditions. The fastest being the other speedy biped with a low power to weight ratio and very efficient respiratory system, the ancestor of the ostrich, an animal which might well have been the pre-human's main prey.

The very features which optimised function as an endurance hunter entailed the consequence of an overall ability to cover far larger territories. A feature which provided the incidental advantages of extending the dietary possibilities by following migration pathways with the seasons so as to realise the many additional possibilities of scavenging, gathering ripe vegetation and also supplementation with fish and marine crustaceans. Already a remarkably versatile bit of machinery

*The Endurance Hunter*

**The absence of a snout:**

An interesting trade-off, arguably the most significant in the light of the subsequent exaptions that were to eventually lead to tertiary tool-use, language and extended imagination, concerns the layout of the head. Other than its particular bipedal locomotion arrangements, the most obvious peculiarity of the human ape is its lack of a snout. Now the snout is a most useful bit of kit for most animals. For the forager it serves well as an extension nozzle and mastication device. For the predator, equipped with powerful ancillary muscles it provides a bite that can seize and kill. For others, such as our fellow apes, it can serve as a very effective weapon for defence or offence. The only real alternative to the bony snout is the very versatile keratin beak, a feature that appears to have been "invented" within the domain of the large dinosaurs, having found a subsequent set of niches only among the birds, so having no precedent in our therapsid-derived lineage.

The poor human, having lost its snout, misses out on all counts. Its browsing capabilities are practically non-existent. Its olfactory tracking potential is reduced by virtue of its usual distance from the ground. We can't use our teeth to catch prey. Despite our belligerence, use as a weapon is usually limited to biting off a chunk of ear. The jaws are not only smaller, but the snout-less configuration renders impractical the implementation of the special jaw-closing musculature attached to the brow-ridge of the skull. Which gives the other apes and various other animals their powerful bite. We can't even us the jaw effectively for carrying. Not that, having hands, we need to.

So why then would natural selection allow an animal to forego something as useful as the well-proven snout? To address this question we will re-examine the requirements for an endurance hunter. We have established remarkably sophisticated thermo-regulation and high power to weight ratio giving lower fuel and water requirements. But we still haven't yet completed the energetic circuit. The other terminal of the battery that accepts electrons. Which for all large animals is oxygen. Of which the long distance runner needs a plentiful ongoing supply.

Now there are often different engineering solutions to a problem. The one used in the racing car for upping the oxygen supply is the supercharger, a device which compresses air before introduction to the combustion chambers of the engine. Among mammals, the other notable long distance runner is the horse. Which, in cooler climates can often narrowly win a marathon contest against a trained human runner. Instead of a supercharger, our equine friend has adaptations that include extra wide air intakes (nostrils) large lungs with high surface area, complemented by a special blood circulatory system that kicks in during intense exercise so as to greatly increase the number of red blood corpuscles to mop up oxygen once it has reached the lungs.

The only animal that consistently outperforms the human is from the family of flightless birds, the ostrich. Its high performance made possible by the superbly efficient constant flow respiratory system used in its avian ancestors. One that originally evolved to provide the very high oxygen uptake rates needed for flight. Together with the far lower avian density, which stems from the many adaptations such as hollow bones and the light-weight beak (derived, like their feathers, from dinosaurs) that enable the majority of birds to fly. Also its own unique set of thermoregulatory features. These latter including bare legs, a long bare neck, vertically oriented at a good height and also the beak. (That the ostrich has this capability to outstrip, at high temperatures, even the human, makes one wonder if these latter adaptations have, in fact, occurred mainly by the selection pressure of pre-human predation)

The solution adopted in the case of the endurance hunting ape was somewhat simpler. Lose the snout and reconfigure the nasopharyx so as to minimize resistance to air-flow while retaining sufficient turbulence to prevent the inhibition of both heat and gas transfers that arise from of laminar flow. Thus allowing efficient breathing through both nose and/or mouth according to demand. (Incredibly, this particular adaptive reconfiguration, which incidentally presents a small additional risk of choking in modern man, is held up triumphantly by many atheistic ideologues as an example of "bad design". Sadly, this group of folk includes a fair proportion of palaeontologists, gene ethologists and many other well-known workers in the sciences.)

Interestingly, the related myth of a descended larynx being a uniquely human feature has now been demolished. A study W. T. Fitch and D. Reby tells us:

> *"The "descended larynx" of adult humans has traditionally been considered unique to our species, representing an adaptation for articulate speech, and debate concerning the position of the larynx in extinct hominids assumes that a lowered larynx is diagnostic of speech and language. Here, we use bioacoustic analyses of vocalizing animals, together with anatomical analyses of functional morphology, to document descended larynges in red and fallow deer. The resting position of the larynx in males of these species is similar to that in humans, and, during roaring, red-deer stags lower the larynx even further, to the sternum. These findings indicate that laryngeal descent is not uniquely human and has evolved at least twice in independent lineages. We suggest that laryngeal descent serves to elongate the vocal tract, allowing callers to exaggerate their perceived body size by decreasing vocal-tract resonant frequencies. Vocal-tract elongation is common in birds and is probably present in additional mammals. Size exaggeration provides a non-linguistic alternative hypothesis for the descent of the larynx in human evolution."*

Which gives support to our contention that this basic configuration was most likely established ***before*** the evolution of complex vocalisations. Perhaps well before. And that the use of speech for language was basically an exaption of a pre-existing hand-based and graphic capability.

The great reduction in bite strength in the pre-human obviously reduced the potential for tearing and chewing meat from a carcass. There were, of course, very nutritious soft tissues such as brain, liver and pancreas to be taken once the kill was opened up. But, particularly for large animals, this may have given similar hassles to those that often frustrate us with today's food packaging.

The solution? Grab a tool, a nice sharp stone to open the skin, a nice blunt stone to crack the skull open. Slicing, mincing and/or pounding reducing even tough muscle tissue to easily digestible size. Similarly for the other aspects of diet, such as tubers or leafy plants, where their often very tough cellulosic structures make for difficult access to the digestible nutrients to be gleaned therein. The increase in specific surface effected in, say, the gorilla by chewing now being achievable by pounding, grinding or slicing. The hands thus taking over the function of the snout, causing the gracile endurance hunting specialist to become an ***obligate*** tool-user. Pushed into a corner by tougher conditions, probably a colder climate, further diversification of behaviours involving more extensive employment of those handy hands in the necessary use of tools would lead this animal along an evolutionary path. One in which its progeny would eventually play host to an entirely new and far more rapid evolutionary phase. That of technology. A path which, short of the occurrence of the most improbable kind of adverse event (catastrophes, of course, follow a power law), would inevitably lead to the kind of ape we are today. A highly imaginative ape. A bearer of language, a fertile soil in which the technosphere could flourish.

**Belligerence :**

A distinguishing behavioural characteristic of *homo sapiens*, as far as we can trace it back, is intra-species warfare. Sure, this occurs within other mammals to some extent, but none of these displays anywhere near the level of belligerence towards its own kind that is displayed by humans. Again, the inadequacies of the fossil record leave us uncertain of the era in which this trait originated. Certainly, the Pan variety of the chimpanzee can be fairly aggressive, but five million years of very different adaptations have occurred since our lineages parted company, so we can't assume the trait to be derived from a common ancestor. Any more than we assume the generally pacific mode of the Bonobo to be derived from that of the Gorilla. From an engineering perspective we can only say that the establishment of this feature provides the positive feedback potential for an "arms race" to flourish.

Particularly in synergy with the development of tertiary tool use. In the light of which we could fairly safely pin it down to the era in which an explosive encephalisation event appears to have taken place. We can usefully categorise the aspects of tool use as follows:

- Basic (primary) tool use is not uncommon in the animal kingdom. Rocks and sticks are used as weapons and for such tasks as cracking open nuts or scooping out insects. Many birds, despite being restricted in terms of manipulation are adept at tool use, the Newfoundland crow being among the more ingenious. Even the building of shelters or nests or hives can quite reasonably be included in this behavioural category although, in such cases we should recognize that the activities are genetically "hard-wired" rather than learned.

  Tool use is greatly facilitated, of course, by the presence of arms bearing manipulative hands rather than paws or claws. A legacy from the primate order as a whole, all members of which have adopted this specialised body-plan feature. The additional innervation required for control of these complex prehensile appendages with extensive tactile feedback being associated with the higher neuron density common to these animals. So it is very reasonable to anticipate an early expansion of that capability in the niche of nomadic scavenger, hunter-gatherer that our Australopith forebears were to occupy.

The human predecessor, however, now an obligatory rather than a habitual tool-user has evolved quite distinctive variations on the primary tool-use theme:

- Secondary tool-use, the use of another tool, which we can look upon as the birth of commerce..

- Tertiary tool-use. The use of a tool to make a different kind of tool. The basis of technology. A self-propagating process by which the generation of different kinds of tools has the

> potential for exponential growth. Not just of the range of artefacts *per se* but of their internal representations, the concepts, the ideas associated with each of them and the relationships between them.

The scene is now set for another phase transition of the overall evolutionary process. Just as with abiogenesis, a radically new niche has become available in the network, enabling a much more rapid further generation of chemical and geometric intricacy. We will deal with this new evolutionary phase, that of ideas and technology in a later chapter. For now, however, to help kill the widely held myth that humankind is not unique in this respect, I quote a passage from one of my earlier works:

> *"The chimp lineage and our own seem to have parted company around five million years ago. Since then our species has come up with such things as the Empire State building, the Eiffel Tower, the Taj Mahal, the Three Gorges dam, the Golden Gate bridge, motor cars, submarines, the hydrogen bomb, heart transplants, tens of thousands of novel chemical compounds, the Sistine chapel, the works of Rembrandt, Turner, Salvador Dali, Shakespeare, Dostoevsky, Steinbeck, Handel, Beethoven, Mozart, Rodin, The Beatles, Charlie Parker, Television, computers, cell phones. We're just scratching the surface, of course. And what, over that same five million years, has the chimp line come up with in the imagination department? A pointed stick!!!. You'd think the moral here would be obvious. But somehow the utterly daft myth of the extraordinarily imaginatory chimp lives on."*

Again, we do not have sufficient information to identify the circumstances that triggered this adaptation but can reasonably assume that it was associated with the encephalization event. Bipedalism haven given the edge on manipulative ability. Locomotory bipedalism allowing teamwork, such as carrying a large kill to camp or a fallen tree to the waterside, a nomadic lifestyle prompting more varied primary tool use such as the gourd for carrying water, the manipulation of fire embers, skins for warmth and protection, the pounding of meat to give greater digestibility, the improvement of weapons for inter-group fighting and for hunting. Any or all of such activities helping to provide a stimulating mental environment favourable to the eventual emergence of third order tool use, the use of tools to make other kinds of tools.

There is another vitally important dimension in secondary and tertiary tool use. For, quite distinct from their primary use, is the place the tool plays as a symbol. A primitive unit of imagination transfer. Indeed, there are classes of tool which perform only that function. Stone artefacts, clearly carefully fashioned, but having no obvious function, may be the first tangible evidence of the transfer of imagination that we call language. Merritt Greenberg and Joseph Ruhlen speculate that these are "rocky" nouns. Even if merely rocky doodles, they still perform the imagination transfer function, as do all subsequent instances of the arts. The scratchings found on a cave wall, too, are of this kind. Their only possible practical use is the transfer of imagination. Made possible by its external storage. The basis of all art as well as the birth of written language. Perhaps the birth of all significant imagination transfer because we are not certain to what extent the vocal channel had evolved in those early times. The synergistic co-evolution of language along with technology providing another source of positive feed-back. Perhaps far stronger even than that of belligerence.

## Socialisation

The other, more widely accepted, accelerant of imagination growth is that of intensive socialization. At first glance this would seem to contradict the belligerence model, but on closer inspection it turns out that these two aspects are not only compatible but actually complementary. Because intra-species warfare is not conducted willy-nilly! It occurs, for the most part, between well-defined groups. Between tribes. And for a tribe to be successful it needs to have strong bonds between its members. Bonds often reinforced by rituals and religious belief. This is a pattern that we see today and throughout recorded history. The tribes of the Australian aborigine, New Zealand Maori, the Amerinds, The clans of Scotland, The tribes of Afghanistan and India. All of these have, until recently, enjoyed butchering and/or enslaving each other as among their most specially favoured pastimes. Interestingly, this close interdependence between group affection and antipathy towards outsiders has recently been found to have a biochemical basis. For experiments using oxytocin, the "love hormone", popularly famous for its rôle in social bonding, have shown that it tends to produce the very reverse effect with respect to strangers.

Where circumstances have been propitious, technology and more ambitious religions have caused the coalescence of these smaller social units into much larger societies. Most of which invariably continued in their bad old ways. And even within these societies we find lesser conflicts, from hill-billy family feuding, through today's street gangs to Mafia wars. The only positive aspect of all this carnage, of which only the broader picture of evolution can make any sense, is that the development of language and the technologies which that enables has progressed in leaps and bounds.

But there can be absolutely no doubt that the tight intra-tribal social bonding in early times provided a great impetus to evolution of language. And that it may have been greatly enhanced greatly enhanced by the use of fire. The warmth and light of a camp-fire furnishing a gathering-place for exchange of imagination long after sunset.

*Consolidation of vocal channel imagination transfer skills.*

This is certainly in accordance with our best estimate of the widespread use of fire immediately preceding the rather sudden increase in neonate brain size of our ancestor that occurred some two million years back. The disparity between head size of the baby and the restricted pelvic aperture of the bipedal mother making childbirth far more difficult and hazardous than in other animals. The head (and brain) of the adult chimp, new-born is far smaller and parturition is no big deal. A chimp's brain is perfect for doing chimp stuff. It does not have to mess around with all the weird imagination-driven behaviours that the poor human ape is stuck with. Like writing this book, for instance. So, for the common ancestor of the human and the Neanderthal (who shared this feature) this is clearly a case in which selection has made a trade-off between child/mother mortality risk and brain size at birth. Presumably to provide the necessary initial hardware for the extension of manipulative capability and imagination required by more complex tool-based life styles. And also, with the reconfigured snoutless enhancement of the vocal communication channel as a bonus, the co-evolution and use of verbal language. This minimal neonatal Basic Input/Output System, unlike that of the chimp or gorilla, being very rapidly supplemented by further neural tissue to meet the greatly increased requirements of the hands and voice together with the additional memory storage needed to accommodate the vast amount of imaginatory software which it will download, Initially, directly from the environment by trial and error, then also, both informally and formally, from the great pool of established human imagination.

It may well be that other hominids, homo Heidelbergis, and perhaps others, shared some of these traits. They developed in parallel with our ancestors only to fall by the wayside whereas we, perhaps merely by virtue of our extreme belligerence, made it through. A common feature of the probabilistic explorative growth of biological and technological networks.

The human animal continues to evolve, and sometimes this takes place quite rapidly. The adaptation of the respiratory oxygen transport mechanism for high altitudes present in Tibetans is reported to have occurred over a mere 3000 years. And there can be no doubt that the easier acquisition of nutrients by cookery and other processing has modified not only our behaviours but also our digestive biochemistries and geometries in generally reductive ways. Or that the protections provided by clothing, shelter, fire-use and the like have reduced some of the selection pressures that would maintain both somatic strength and resistance to cold.

But the environmental changes wrought by technology now change human behaviour, and some aspects of its form, far faster than genetic mechanisms.

We like to hang on to the cosy notion that humans are the creators of technologies. A closer analysis of the evidence shows this is not so. In fact, it is perhaps fairer to say that the modern human is a product of technology.

# 24

# THE MYSTERIOUS STRANGER

*At a very early stage in human pre-history, a passenger unobtrusively boarded our evolutionary train.*

*A fellow-traveller who would prove very helpful on our journey into the future and who would eventually determine the path to its present destination.*

*An indispensable companion whose offspring would one day thrive in the language-rich soil of our collective imagination.*

*The name of this mysterious stranger with whom we have co-evolved is Technology. Our species, as a whole, is no less separable from her than is the chicken from the egg. For the present, at least, we are mutually interdependent.*

*She has helped us through ice ages and enabled our position at the top of the predatory chain, producers of a continuous supply of fine nutrients by farming.*

*As well as significant protection from weather and predation at all levels.*

*She has enabled the distribution of humankind throughout all four corners of the world.*

*Her tender mercies have supported the explosive growth of our population. And, in her darker moments, provided the means for our murderous instincts to substantially destroy whole swathes of that population.*

*She has provided all manner of arts and diversions and to her own evolutionary advantage, has endowed us with an appreciable understanding of nature's machinery.*

*For the vast majority of us her allure and her fascination is irresistible.*

*We shall meet this lady in chapter 29*

# 25

# THE MYTHICAL"BRAINY" APE

There is perhaps more secular myth associated with the central nervous system of humans than any other biological entity. This applies even to many of the professionals whose occupation brings the "brain" within their chosen field of study. As well as the general population, most of whom give it little consideration. The myths arise from blind acceptance of tradition rather than objective evaluation of prevailing evidence from a sensible engineering perspective. An example of such a myth, proclaimed even by many who would deny human exceptionalism, runs along the lines that "The human brain is the most complex entity in the universe". Which is patently absurd, if only for the reason that this particular network is a small subset of the wider network represented by collective human imagination, itself a subset of the network which is the biosphere, this again a subset of our solar system, itself of the galaxy.

At a more mundane level, can we with certainty affirm that it is a more complex construction than of that awe-inspiring masterpiece of chemical engineering, the human liver? Or than the brain of an elephant? Or, even today, the network of the Internet and its peripherals? Some of these latter systems are extremely difficult to accurately quantify but they are certainly contenders for the distinction. Another nail in the coffin of the "superbrain myth" comes from recent studies which indicate that the total number of neurons in the human brain are as predicted by the almost linear allometric scaling within primates.

Another common myth comprises the almost universally accepted but completely unwarranted assumption that our use of mathematics is the pinnacle of "intelligence". Chess players, too, are commonly considered as exceptionally "intelligent". That such beliefs are widespread is amazing. Playing the game of chess is one aspect of human imagination that has long been surpassed by individual computers, far before the present state of the art. Compared with the computations required by real-life physical activities, chess is trivial. By way of comparison, the game of snooker, as practised by a professional is far more demanding. And I suspect that, even today, any individual computer would be hard pressed to effect a win against a player like Ronnie O'Sullivan. And as for that enormously useful tool, that great utiliser of simplicity, the specialized subset of language which we call mathematics, we find that it is largely a retrograde exaption of the lightning fast computations that the central and peripheral nervous systems (we should not really make a distinction in this context) use routinely in the course of their mostly subconscious duties.

Yet more myth, this time involving a sin of omission rather than assertion, is the vague notion that computational capability somehow represents the be-all and end-all of "intelligence". It is very easily demonstrated that this is most certainly not the case. While it is experimentally possible to keep decerebrate animals alive, it is inconceivable that a child born with no cerebrum and cerebellum could ever be able to learn language or play tennis. The computational capabilities for such behaviours would clearly not be available. Furthermore, a child having a normal brain but deprived of all sensory inputs would quite obviously not be able to acquire these behaviours. Computation alone is not enough. Equally, or perhaps more important are input and output devices. Ronnie would win no snooker tournaments wearing a blindfold or with his arms strapped to his sides. Nor would Serena Williams succeed at tennis. Such things seem rather obvious but, particularly when considering non-human computers and the Internet, the crucial importance of input sensors and output effectors is usually overlooked in estimating levels of "intelligence" or, more sensibly labelled, imagination.

So if we are to abandon the dearly cherished notion that, even if not "The Chosen People", we are at least deserving of being intra-cranial possessors of "The Most Complex Entity In The Entire Universe" or particularly deserved of the title "The Brainy Ape", to what, then, can we ascribe the very atypical behaviours exhibited by the human animal who uniquely seems to have the knack of generating billions of artefacts and systems? Or of generating entirely new niches to occupy? As well as, by way of spin-off, engaging in myriad bizarre behaviours which are in no way directly related to replication or survival? Strictly speaking, there is no biocentric escape from this enigma, but for the time being, in the interests of softening the blow and retaining a little our dignity we will make our claim to fame. Our special feature, language, together with the greatly increased memory capacity that is required for it. But this is not a mystical property to which all other organisms aspire. No! In essence, it just boils down to hypertrophy of those parts of the nervous system particularly involved with the operation of the hands and vocal apparatus.

For, while the other apes continued to make a hand to mouth existence, those of our lineage, moving into the niche of endurance hunter, increasingly by exaption transferred much of the function of the snout to the hands. Of necessity increasing the further use of tools both for tearing flesh and comminuting and tenderizing by pounding. And by exaption of this largely obligate behaviour, learning other ways to use tools, notably to make other tools, securing a selective advantage. Both for pre-human evolution and a new evolutionary phase, that of technology. Setting up a nice feed-back loop which was to lead to the co-evolution of language together with its complement, a greatly enhanced imagination

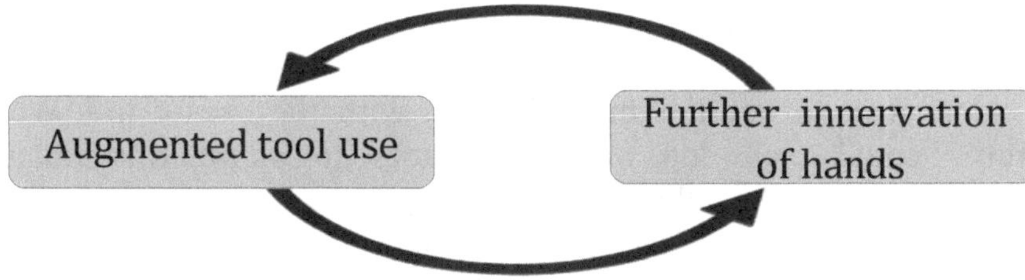

The physical structure of the hand in itself does not tell the whole story, the progressive sensory and motor innervation via the CNS to provide increased control is of even greater importance. Nor does manual dexterity in itself give us the full picture.. The handy human ancestor may well have been pre-empted by some kinds of monkey. Certainly there are some of these today who are very adept. The Macaque, for instance, for whom, for some reason, most likely the fact that they and they and their ancestors were not snoutless. The amplification of tool use cycle, in this case not being obligate for nutrition, had insufficient selective advantage for it kick in.

Once established, the co-evolution of technology with the human involved not only far greater manual dexterity but also the selection pressure for increased imagination to cope with the increasing range of uses and skills. So we have an accelerating and self-sustaining concurrent increase in manual dexterity and imagination. We can choose to throw in here another positive feed-back amplification. That of intra-species warfare. This appears to have been a distinctive feature of our species for a long time. There are occasional examples of this in other animals. Attacks within the chimp world, for instance, do occur. And there are a few peaceable human groups, of which the bushmen of the Kalahari are an example. But societies such as these are exceptions rather than the rule. Whereas, for humans in general, belligerence seems to be very much the rule. In such a scenario, we can see that fighting an increasingly weapons-equipped adversary of similar cunning could well be expected to provide a further selection pressure for imagination and the neural hardware in which it runs. Also, perhaps, for the enhancement of the vocal communication channel. A small start to the great sophistication of this apparatus which, already physically advantaged by the reconfiguration required by the snout to hand transition, was to blossom later.

We come next to another breakthrough with the use of fire. What features could have led our progenitors to adopt this unique behaviour? It is generally believed that the acquisition of embers or burning branches was the first resource and the serendipitous combination of sparks from knapping with dried plant dust or by attempts at drilling timber came later. Giving a less formidable source. There seems to be no strong evidence either way. In either case, bipedalism, with the hands freed up for other uses, would clearly be advantageous. Another factor could be that our nakedness outweighed the animal's usual fear of fire by providing warmth on cold nights. Certainly the ever-growing use of imagination resulting from the earliest pre-stone age wood, skin, fibre and bone technologies could have helped. With combustion under control, the scene was set for the rapid development of the large brained hominids of which our kind is the remaining branch. The control of fire opening new vistas in chemistry, particularly the art of food preparation. Providing greater bio-availability of nutrients. Allowing less investment in gut and greater investment in innervation of hands and mouth and imagination. Giving greatly increased survivabilities through the rigours of the first Paleozoic ice age, together with more intensive socialization around its warmth that would stimulate imagination transfer. A fertile soil in which both spoken and graphic language could blossom.

So it came to pass that around the time of the second ice age of the Paleozoic era, the large-scale implementation of the hand and mouth specialized ape emerged. Complete with the adversely oversize neonatal head required to accommodate the additional control systems in its early stages of development. The replicative advantage of these imagination-transfer systems enabling features that well outweighed the additional risks incurred in child-birth. As well as the investment in protracted maternal dependence. For the snoutless bipedal ape with the fancy hand and mouth systems, although still by no means having entirely lost the capability for endurance hunting, was now born with the basic mind-sharing neural equipment that would make possible the great flowering of technology that was to come.

Studies have been undertaken in recent years using an isotropic fractionator to derive accurate cell counts for major regions of primate brains that have been homogenised. Fluorescent labelling of the DNA indirectly allows an accurate neuron count. This work has shown that the neuron density of primates is much higher than that in other orders, such as rodents or insectivores. Also that, within the primates (though not within the other orders), the total (and component) brain neuron count scales in a fairly linear fashion against size. Also that the number of neurons in the ***human*** brain is as predicted by this scaling.

Quite apart from brain size, neuron density and the very difficult issue of connectivity, of which our understanding is very far from complete, there is the fact that the metabolic rate of the human brain is inordinately higher than that of other primates. This extra energy expenditure suggests either more extensive or faster processing of information. Probably enabled by exaption of the additional cooling of the brain that had been provided by courtesy of our historic adaptation to endurance hunting. These extra heat exchange features are not present in the other primates. The most distinctive feature of these features is, as pointed out previously, a reticulum of veins in the skin of the head in which the direction of blood flow is ***reversible***. such that the cooled blood can flow back from the skin surface into the brain. An exaption that copes well with the new demands of the increased level of information processing work that is required for the complex control of hands, vocal apparatus and their conjugates, greatly enhanced imagination and language.

It has long been one of my pet beefs that the possibility that the cerebellum might in fact make an important contribution to the property we call "cognitive ability" has been neglected. This feature having usually, in the past, been exclusively attributed to the cerebral cortex. But the long underestimated cerebellum has indeed been found to have greater raw processing capability than the cortex, and, what's more, by a very large margin. Frederico Azevedo *et al*, using the homogenisate method mentioned previously, have found that the neurons in the cerebellum actually exceed those present in the cerebral cortex by a ***factor of four***. A proportion, incidentally, which applies approximately across the spectrum of other primates. Thus, it would seem that the somewhat speculative scenario portrayed in my first book, "Unusual Perspectives", whereby the most intensive computation perhaps takes place not in the cortex, but instead, in the cerebellum, has now been vindicated, The cortex being primarily involved in some of the rather more leisurely activities related to navigation and archive maintenance. In other words the processing and distribution of sensory inputs and the handling of an unusually large amount of memory. Tasks most closely associated with imagination.

So we see that our celebrated "cognitive abilities", which after all, can only be objectively assessed by observing behaviour, essentially resolve to the primarily cerebellar activities that concern the manual and vocal channels. The particular innervations that enable the function of language. The ***sharing*** of rather large imaginations. And that the "Brainy ape" is not really all that brainy, and may actually turn out be, in terms of neurons, rather less brainy than some other species. That long-held illusion is shattered. But we have come, I hope, to realize that this is really no big deal. All organisms, *staph. aureus*, the mushroom, the bee, the lily, the star-nosed mole, the sheep, the octopus, the gorilla and the human, all have their area of specialisation. All excel at what they do. The layout and organisation of their nervous systems are task-oriented. Each being optimised , as far as is possible, for the its efficient interactions with its own niche. Interactions within the swarm of cells that comprise the individual, and interactions with its external environment. After all, that is what nervous systems are for. Each kind of organism, after allowing for the redundancy and spin-off that bestow resilience to the network, plays its part in the evolution of the biosphere. The niche of the human, our own particular area of specialisation, just happens to be the implementation of tertiary tool use and language. For which our bodily forms and central nervous systems are uniquely configured. It is these two complementary features that have led to the inception of a radically new departure in the overall evolutionary process: The technosphere.

# 26

# THE SMOKING GUN

We have come a long way in our historical journey, having tracked an indisputable directionality right through from the formation of the chemical elements in stars and supernovae to the technological world in which we find ourselves today.

The evolution of a network of ever-increasing chemical and geometric intricacy. A process so persistent and pervasive that it not to be reasonably ignored or, without evidence, swept under the carpet by invoking scenarios grossly lacking in parsimony. Such as an infinite number of universes, each endowed with different fundamental physical properties. Or the purely anthropic notion of a "designer". Especially as on close analysis we find, even at the human level from which the concept is derived, there no evidence of a "design" process that cannot more accurately be attributed to evolution of ideas.

The body of evidence for the various phases of the evolutionary process varies. For the formation of elements it is quite strong because we are able to observe stars in all stages of development and determine their metallicity spectroscopically. Effectively taking a snapshot of history that gives a coherent and generally accepted picture of this first phase.

The evidence for the mechanism of the formation of solar

systems and of galaxies is more sketchy, but the very presence of these net-worked structures, in conjunction with the implications of nucleosynthesis provides strong support for increasing intricacy.

Similarly, the increase of geologic chemical intricacy of our particular planet, while partly speculative, has a reasonably sound empirical basis. And there is no good reason to suppose that similar processes would not occur on comparable planets in some other solar systems.

In contrast to the previous two phases, the evidence for the immense overall increase of chemical and geometric intricacy in the biological phase is overwhelming. As is the extreme level of directionality imposed by the well-established ratcheting process of natural selection.

But it is in the most recent phase, the evolution of technology within the shared minds of humans, that the evidence is most accessible because it is in the domain of both recorded history and the present.

It would be very surprising to find that, having come all this way through the aeons, the increase in intricacy, enabled by so many of its components being consistently "just right", should suddenly, in its technological phase, have come to a grinding halt a few hundred years ago.

And that, in the absence of at least three materials being available and "just right" is exactly what would have happened.

For me, their properties most convincingly represent the "smoking gun".

We will take up the evolution of technology in chapter 29

# 27

# CONSCIOUSNESS IS NO MYSTERY

In any discussion of the overall directionality of natural process the question of consciousness inevitably arises. This has traditionally been considered to be a phenomenon uniquely associated with the human animal and with the mythical concept of a "soul". Religious or mystical beliefs to which, even in these more enlightened times, the majority of folk still cling. Sadly, such convictions are, even now, endemic among many of those who are working within, educated in, or are associated with the sciences. From the hypothesized universal property of "dark energy" right down to quantum effects in the microtubules in the cell, bizarre speculations of this kind are always forthcoming.

The intractability of drawing a circuit/flow diagram to map the biochemistry and geometry of an organ evolved over the course of billions of years is unquestioned. The same applies just as much to such organs as the liver or the pancreas. The level of complexity of a single cell alone is prodigious! But that is no justification for resorting to metaphysical interpretations of these mechanisms.

Furthermore, within a quite different context, that of biological evolution by natural selection, the nature of consciousness (self-awareness) is certainly a mystery no more. So, to sensibly get a handle on consciousness, we must look not to the enormous complexity of neurology, but elsewhere. Finding that we have at our disposal today conceptual tools that provide a full empirical understanding of the ***general nature*** of consciousness. Firstly, and most importantly, from our understanding of biological evolution by natural selection it becomes quite clear that the provision of a navigational feature that involves some degree of self awareness is required for an organism to interact optimally with its environment. It is a measure of its fitness for the prevailing environment and subject to selection pressure accordingly. There is, of course, a great gulf between the level of consciousness exhibited by our species in comparison to any other. Simply because the level of interaction with the environment required by our particular ecological niche is incomparably greater. As we have seen, this is evidenced by the billions of artefacts and systems that have resulted from human activities. Secondly, we can be quite sure that consciousness has a purely chemical basis by virtue of the fact that it can be easily "turned off" chemically by means of anaesthetics. What's more, both the state and the perception of consciousness can be modified at sub-anaesthetic levels by such substances as diethyl ether, and nitrous oxide. Not to mention the very wide variety of altered states of consciousness induced, again chemically, by substances such as alcohol, cannabis, amphetamines, cocaine, LSD and so forth. Thirdly, from computer science, we now have an exhaustive understanding of how, at the nuts and bolts level, information processing systems work. While neural mechanisms are very different from the digital electronic circuitry of our computers, there are sound functional comparisons to be made and to help our understanding at a general level.

We must always bear in mind that, for the most part, our central nervous system performs its multitudinous complex tasks without any awareness on our part. The consciousness is merely a tiny window on the world of which we are part. Essentially just a navigational facility. Albeit a rather nifty and important one. Perhaps the most serious barrier to a sensible comprehension of consciousness and related phenomena is attributable to the unfortunate use of the vague and silly word "intelligence". This blighted term, with its huge sliding scale of connotations, many of them emotive and almost all of them anthropocentric, is responsible for much of the woolly thinking that prevails in discussions of this kind.

The particular feature in which our particular species excels can be better described as "imagination". Defined as "the ability to construct, store and morph neural representations of the external world within the mind", it corresponds quite closely to the everyday use of the word and is relatively unambiguous. It is a feature that gives a strong survival/replication advantage for most situations so it is not at all surprising that selection pressure has effectively installed various levels of imagination (or its precursors) in most organisms.

Consciousness represents an awareness of self, even at the most fundamental level. The requirement for an active sensory information input defines a sensible boundary that will exclude such things as rocks, socks and brooms. Concisely, we can say that:

**Imagination = Consciousness + Memory + Morphing**

*Where memory is read/write information storage and morphing equates to shuffling and recombination of ideas.*

So, while it may be that a bacterium has only consciousness, the slime mould, so it would appear, also has some primitive form of memory. Whereas the cat, for example, has all the components of imagination.

But natural selection has honed the imagination of our particular species to a prodigious level. While the "mind" of such things as the slime mould would appear to have a relatively simple biochemical basis, the advantages provided by features of this kind have, in many organisms, most notably the mammals, been implemented with ever increasing levels of sophistication in the form of the specialized electrochemical nervous systems with which we are so familiar. There is no clear line that can be drawn! Most, if not all organisms must, in principle, have a certain degree of consciousness (self-awareness). Even if only as the locus of its sensory and effector interactions with the external world. This certainly includes the simple tropisms observed in such creatures as bacteria and plants. Tropisms from which, ultimately, imaginations such as ours have evolved.

Again, these essentially navigational features are, in the case of more highly organised creatures such as the vertebrates, far better described as "imagination" rather than by the use of the horrendous discombobulation of "intelligence". Imagination, in various species, particularly our language-enhanced one, has been exapted for diverse tasks which, at first sight, may not appear to involve navigation. But closer examination shows it to be so.

We can define memory as comprising the ability to store, in a retrievable form, information derived both from the external environment and, in many creatures, also from internal sensors which provide status information. Within this context, memory can be likened to that componentry of a computer that holds software, in contrast to the hard-wired information that constitutes the organism's hereditary basic operating system. Morphing, a word I have stolen from computer geeks, is the ability to merge, meld, reconfigure and diversify the neural constructs that are embodied in memory. It is very important to remember that mind is ***not*** the boss of the body.

The ancient belief that the mind is in some way the master of the body should really, in the light of modern science, be long dead. It is now very clear that the mind is simply a function of the community of cells which the body comprises. Also that the rather small aspect of mind which we identify as consciousness is merely an extension and elaboration of the navigational facility required by many organisms for optimal interaction with their external environments. The actual extent of such interaction of our own species, of course, extends far beyond the simple requirements of food, sex, basic shelter and defence against predation that are adequate for many animals. Accordingly, the extent of consciousness (enormously enhanced by our sharing of imagination by means of language) is, of necessity, correspondingly greater. The mind is the ***servant*** of the community of cells that comprise the body, rather than the master.

From time to time we hear of the "discovery" of "the seat of the mind". The precuneus and temporal parietal junctions are some recently promoted examples. While these regions are important components of the overall mechanism of consciousness we must be careful not to identify features such as these with the phenomenon itself, or to fall into the trap of assuming that consciousness somehow "resides" there. As with the "qualia", the best interpretations of modern science strongly indicate that the various levels of consciousness are distributed functions rather than being confined to specific locations within the brain. That they are a network function!

Certainly, with our present background of understandings drawn from the main body of science, it is hard to envisage that this would not be the case. The results of experiments using fMRI, PET scans, and so forth, while valuable and intriguing investigative tools, are subject to misinterpretations. Because it is pretty safe to say that brain activity is essentially characterized by switching operations. Also that the switching can occur at local or remote points. So, basically, all we see when fMRI detects higher blood flow or PET detects higher glucose utilisation, for instance, is a greater switching activity within a particular location. It is quite wrong, however, to conclude that the processes involved are confined to, or even predominate in that area. The only safe observation is that the area that "lights up" is a network hub. To make this point clearer we can use the everyday example of a city rail network such as the London underground. If we zoom into the central city area we detect a far greater level of rail movements than if we look at similar sized areas in the suburbs. Little green men (rather naive ones) making these observations might conclude that the "seat of rail transport" lies at the city centre. Whereas rail transport is actually a function of the whole network. I make these points as a side-issue. The detailed operation of the brain is far from being well-

understood and it is my personal opinion that it is very likely that it may ever remain so. That, in that particular sense only, simply because of its great intricacy, it may very well stay a mystery.
Even within reputably secular circles, crackpot theories that are essentially mystical or metaphysical still abound. It is amazing that so many workers in the sciences become tied up in knots with what, in the light of present knowledge, is an easily explained phenomenon. In their attempts to cope with the mystical overtones generated by the traditional introspective approaches to the subject, with the consequent recursivities, all manner of quite unrelated or pseudo-scientific concepts are brought to bear. Often invoked to explain what is really a non-issue are such categories as "emergent dynamics", information theory and autocatalysis. Phenomena blithely and quite spuriously invoked by others include quantum effects, fractals and "complexity theory"

Salient examples are Erik Andrus(PhD, Assistant professor of molecular biology, author of the peer-reviewed "Complete and Consistent Theory of the Universe" who claims that:

> *"Basically, the complete theory (aka theory of everything) proves that consciousness is dark energy. And that what anyone calls, 'My Consciousness' is My Consciousness. ~74% of the Universe."*

Theoretical physicists in particular are far from immune from such silliness. In a somewhat similar vein, the platonic professor Paul Davies assures us that the evolution of humans somehow gave our whole universe self-awareness!

And we have the offerings of the well-known Hameroff-Penrose double act: with their "theory of consciousness". Which, despite frequent mass media claims, has absolutely no evidential basis and is a speculation which, again, can fairly be described as a mere exercise in quasi-religious pseudoscience. As a correspondent remarked:

> *This 'quantum consciousness' idea generally revolves around the absurd syllogism that because quantum mechanics is a mystery, and consciousness is also (to H&P) a mystery, therefore the two are linked."*

Having said that, we can of course have absolutely no doubt at all that quantum effects underlie ***all*** chemical systems, including the large subset represented by biology. That, however, does not make them agents of consciousness, a phenomenon which, as we see, is very easily accounted for within the macroscopic world of well established science. Again, it is important to remember that the understanding of the nature of consciousness does not lie in elucidation of the precise mechanisms of the CNS. While substantial advances have been made in the determination of gross structure and function of the nervous system as well as much of the underlying biochemistry, it is still very much on the books that we will never fully unravel this exquisitely complex system. The answers to "how" and "why" are not to be found by that route, any more than are the detailed function and "purpose" of, say, the liver. The precise nano-scale mechanisms of that organ, too, are well beyond our present comprehension but we do not hear pronouncements concerning the "mystery of hepatic chemical processing" or the attribution of mystical interpretations to it.

There is an analogy with, say, a motor car. We can reasonably say that, in functional terms, we might know how drive it, why it exists (its "purpose") and how it evolved, all without having any knowledge of what goes on under the bonnet. So, to sensibly get a handle on consciousness, we look not to neurology, but elsewhere, and, as we have seen, find that we have at our disposal today conceptual tools that provide a full empirical understanding of the general nature of consciousness. That consciousness is simply a navigational tool, a user interface to the external world of which we, too, are part. Such a realization may, of course, offend anthropocentric conceits. The mystical interpretations are, today, redundant. It has, for some time now, been fully apparent that our unusually high degree of self-awareness is a rather straightforward feature that has been thrust upon us by natural selection. This soundly established process of biological evolution by natural selection has given us considerable insight as to how structures arise in creatures and of the reasons that underlie their function. With the wealth of understanding of our world that we have today, we are fortunately no longer hide-bound by the inevitable self-referential recursivities of introspection which, in the past, generated the mystical associations that found a ready home in religious mythologies. "What is consciousness" when stripped of this spurious metaphysical baggage, is seen to be rather easily accounted for on an entirely empirical basis. In fact, it is, as we see, an evolutionary necessity. Again, simply that navigational facility which enables an organism to interact optimally with its environment. The level of interaction of our species being inordinately high and the extent of consciousness being correspondingly sophisticated.

The important distinction here is that the consciousness, self-awareness and, for that matter, the free will that is exhibited by humankind differs from that of other organisms quantitatively rather than qualitatively. There is, of course, a humongous gulf between the essentially navigational feature exhibited by our species and any other. But all creatures require such a facility to allow optimal interaction with their environment. Language is most certainly the crucial component and *raison d'etre* of our own particular level of consciousness. However it does not fully define it. Would we say, for example, blind, deaf, mute Helen Keller was not at all conscious or self aware before acquisition of linguistic skills via the tactile route?

It is often claimed that there is a profound difference between how a computer operates and the function of a mind. That a computer can process information very quickly, but it is not creative, it simply follows rules it has been programmed to follow. That it is incapable of making creative choices. That consciousness is exceedingly more complex than that, and emotions, feelings and meaningfulness are all essential components. Claims such as this merely conflate the level of consciousness with its existence. These things are simply a matter of degree, of the level of functional intricacy. As discussed in a previous chapter, it is only in a trivial sense that the human mind is creative. On closer examination it is found that "creations" are illusory, that they are but part of the overall evolutionary process.

Emotions and feelings are inevitable features which have arisen by natural selection to optimize the fit of an animal to its niche. Certainly emotions are essential but they are, at root, task-oriented and tailored to a creature's requirements. Those the human animal has at present, however, are uncharacteristic in two important ways. Firstly, owing to the recent very rapid change in its niche resulting from the rapid advance in technology, emotions associated with belligerence are now extremely sub-optimal. Secondly, because of the transformative emergence of collective imagination enabled by language, the human mind has now a new dimension of consciousness. It is not only self aware but is ***aware*** of being self aware. Aware of awareness! Rather than merely engaging in behaviours driven by the biochemistry of emotions it also ***reflects*** upon those emotions. Indeed, some of us even reflect upon the biochemical mechanisms themselves. And others, using this unique retrospective capability less productively, all too easily becoming sucked down into self-referential whirlpools which give rise to the illusions of spirit and soul.

A computer equipped with a web-cam pointing at itself is, in the most primitive sense, self-aware. As is a bacterium. Any system provided with some sort of sensory input about itself must thus be conscious to some degree. We cannot, of course, expect that minuscule level of consciousness to in any way resemble that of our species, with its millions of photo-, tactile, proprio-, thermo-, chemo- and other receptors that enable it to self-identify. As well as the wonderful complex limbic and hormonal systems that have selectively evolved to provide complex emotions. All in addition to the exquisitely sophisticated neural information processing mechanism which manifests itself as imagination. Not to mention the important contribution to "understanding" that corresponds to the ability to export, import and externally store imagination. The feature that we call "language" A further consideration to be taken into account is that, for this self-awareness to be experimentally demonstrable in terms of behaviour, the system needs to have some kind of basic signal processing system as well as a motor output. Exemplified by the many simple robots available today which autonomously perform tasks such as materials handling. Often capable of, when hungry (sensing a need of high energy electrons), navigating their way to and docking with a charger. Such devices routinely demonstrate a certain level of self-awareness. And, indeed, of self in relation to aspects of their environment. Sure, they are pre-programmed to do such things. But so are we. By natural selection, by copying, by training and by acquisition of data in the form of language.

At the other end of the scale we find ourselves. The human animal. A hugely intricate network of cells that employs an extremely complicated navigation and control system which we can loosely identify as the "brain". A system that has many levels of awareness. Awareness and control of such essentials as heartbeat, breathing, intestinal peristalsis and other autonomic functions. Awareness of electron deficiency (hunger) or, say, bladder or colon fullness. These aspects of consciousness running in the background unless situations arise calling for their promotion to the attention of the primary navigator. The awareness and control systems relating to positions of the limbs, too, quietly operate in the back-ground most of the time, even when active, For example, the primary navigator, having been alerted to thirst by a sub-system, can set the muscles to walk mode and leave them to function in the background while it uses sensory inputs and/or memory to guide the swarm to a source of water. In fact the very capable modern navigator, while driving a car, can even outsource that task to a lower level autopilot, who will take over the relevant sensory inputs and the motor outputs while the primary navigator, language enabled, ponders such abstruse things as, say, the nature of consciousness.

For that is a very special feature of the human ape. Because of the unusually intricate imagination that has co-evolved with the unique network of language, this creature is, very strangely, not only conscious, but conscious of being conscious. Which, if we are not very careful, can easily allow our cogitations on the subject to become locked into introspective loops. Oh to be a pussy-cat, a dog or a frog, with no such worries!

# 28

# *QUO VADIS?*

*As discussed in chapter 27, it is entirely proper to attribute some level of consciousness to any device that has a sensory input and a data processing module that selects an appropriate effector output. Because natural phenomena arise incrementally there are no sharp dividing lines to be drawn. Just as with the transition from pre-biological chemistry to biology. In the present case it would seem sensible to arbitrarily set the most primitive consciousness somewhere between the thermostat and the Roomba.*

*At the other extreme of human-mediated intricacy we have the Internet. This rapidly evolving construct has a surfeit of sensory inputs and effector outputs, with immense amounts of processing capability in between. It is thus endowed with a vastly greater level of consciousness than a Roomba. Far higher even than that of a bacterium or a plant or a sponge. In fact, more than most mammals.*

*In terms of hardware, measured by the number of basic computing elements (semiconductor switches) it may already surpass that of the human.*
*But sheer weight of processing elements alone is not sufficient to match imaginations comparable ours.*

*For this, the ways in which the parts are interlinked are important. And also the extent of memory. There is no problem with the latter - the Internet has an abundance of memory. Memory that is even now well replete with the sum of human knowledge.*

*The topologies of its interlinkages do not yet compare to ours, although the patterns of growth strongly suggest that in many respects they will soon do so.*

*By extrapolation of the exponentially growing techno-evolutionary trend which has been occurring over the past two million years we can confidently predict that the Internet will soon become the new predominant cognitive entity on this planet.*

*Also that, since the prospects for an inorganic "life-form" on this nursery planet would seem to be very limited, its progeny will almost certainly be adapted to an extra-terrestrial niche.*

*We will take a speculative sneak preview in chapter 31*

# 29

# THE EVOLUTION OF TECHNOLOGY

We have noted in chapter nine that our commonplace notions concerning design, creation and invention are key components of a paradigm that, like that of Newtonian mechanics or the Bohr model of the atom, is fine for many humdrum purposes but entirely inadequate for any serious discussion of natural processes. Just as we now recognize that biological organisms do not spring into existence as the result of the activities of a "supernatural" being. Or, according to the belief prevalent in the nineteenth century with regard to bacteria, yeasts and moulds, by spontaneous generation. It is very apparent, by careful objective analysis of history, that ideas and the artefacts which embody them are products of an intricate network which grows incrementally. It is not only by reference to history from which this interpretation is drawn. In what is perhaps the most firmly established branch of science, that for which the accrued evidential basis is strongest and the applications most far-reaching, the science of chemistry, there are to be found no true discontinuities. Reactions, though usually having a strong stochastic basis occur incrementally, and, in cases where they are sufficiently well understood, very predictably.

In, by way of example, the Fischer synthesis of ethyl acetate, the yield under specified conditions is found to be constant and repeatable. It is entirely predetermined. But it is in a state of dynamic equilibrium. Individual bonds are breaking and reforming all the time. Very importantly, which particular atoms will wind up in the product is entirely unpredictable. Most certainly not predetermined. This is an example of how nature's machinery seems to work. Gross outcomes of stochastic processes are often predetermined. The activities of individual nanoscale entities being random. But probability nevertheless determining the final concentration. Also, we find, in the case of abiogenesis, the dividing line between chemistry and biology is fuzzy, artificial and arbitrary. Similarly, because of the incremental nature of network growth, so it is with atechnogenesis.

For present purposes we will pick up the roots at that time in pre-history when a certain ancestral animal had acquired the prerequisite adaptations of terrestrial habitat, very efficient thermo-regulation, obligate bipedalism and a significant measure of manual dexterity. We can reasonably set this time as somewhere between 3 and 4 million years back. Well after the time of the divergence of the chimp and gorilla lines from our common ancestor. Somewhere in this period we can place the further adaptation of obligate tool-use. Since the earliest explorations of such technologies almost certainly used organic materials rather than stone they have succumbed to the ravages of time, leaving little chance now of finding evidence for them.

Timothy Taylor, in "The Artificial Ape", ingeniously suggests the baby sling as a likely early candidate, such slings being used also for carrying food and other items by gatherers even today. Others are sticks, for prying and, in conjunction with skins, for building shelters such as wig-wams, and the use of skins and feathers for clothing. The use of a floating log for water transport, or the gourd for carrying water – not an option for the quadruped, nor an advantage for species not of a nomadic bent. Also, well before the "stone age", we can quite confidently guess that these handy hands and the more gracile arms to which they were attached would have been have been further employed in the use of unworked stone as projectiles for hunting. Another adaptive feature providing selection for increased specialised neural configurations of the fore-limbs and computational power for their control.

This regular employment of handy hands providing a selective advantage for the further adaptation of increasingly handy hands. Providing a positive feedback loop that may very easily have found the trick of making a primitive knot. Chimpanzees have been observed to make "accidental" knots that often hold, and the Weaver bird can tie true knots. (Admittedly, both of these lines have had hundreds of million of years to brush up their skills since the time of these human ancestors.) Toying with these ideas I happened upon a remarkably secure and simple knot-less fastener using just flax fibre and two sticks.

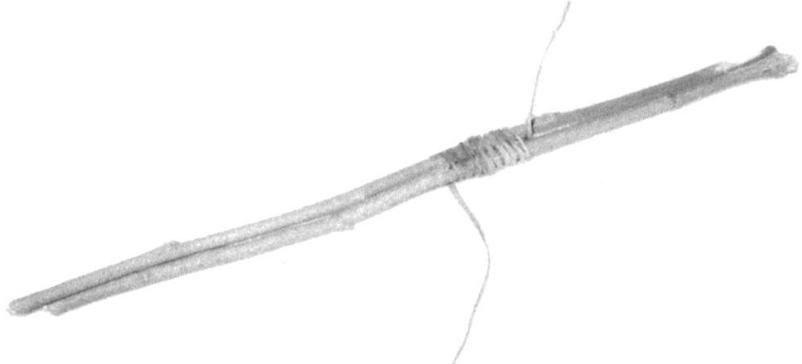

Using gut, vegetable fibre or sinews, even our pre-stone age ancestors may well have reached the threshold of tertiary tool use by making composite structures thus fastened. Even awls or primitive needles or bodkins fashioned from wood, bone or plant spines can not be ruled out as a means of production of composites. Improved fasteners opening the door to such things as rafts, more effective shelters and clothing, perhaps even crude baskets. Whenever technologies such as these emerge, they have the potential to provide a selective advantage for the user and thus, by proxy, to themselves. Thus, when technological innovations give some kind of edge under the prevailing conditions a positive feedback loop is set up that will tend to accelerate their replication. So, by the time that the fashioning of stone evolved about three million years ago it is very reasonable to assume that these predominantly nomadic endurance hunting ancestors were significantly immersed in proto-technologies, the development of which greatly enhanced the novel use of efficient cutting tools. At some time in the next two million years or so, fire was befriended, The bipedal configuration facilitating the carrying of burning branches or of embers for re-ignition. At a later time the initiation of fire. Perhaps simply by the chance encounter of sparks produced by knapping flint with dry organic scrapings. Or by the generation of heat by primitive drilling endeavours.

Controlled fire leading to the serendipitous realization of cooking. The first chemical art. One that, by reducing the time and energy involved in the seeking of sufficiently nutritious food sources, enabled greater investment in behaviours that were to favour brain development. An advantage to be later reinforced similarly by the adoption of agriculture. Also, equally surely, given the human propensity for playing with fire, the eventually gleaning of metals from their ores as well as their thermally assisted fashioning. The pace of technological diversification, its evolution now further fuelled by fire, greatly outstripping that of its host, the pre-human animal.

Not far down its serendipitous path was the emergence of the rudiments of another kind of material which, many centuries later, prove absolutely crucial for its own continuance. The alkaline metal silicates which we generically label as glass. Some other kinds of glass, the obsidians, had been around for aeons. Indeed, these silicon dioxide based minerals of volcanic origin had been used in the stone ages. Used by virtue of the extreme sharpness of their fractured edges. Not for the key features which would make glass so vitally important in times to come. Among which were ease of working and optical properties.

Most likely the development of cookery, and the much greater range of plant foods thus made edible, accounted for the shift of human foraging away from the hunter to hunter-gatherer. The continued symbiotic evolution of technology and human society producing a new geographic niche to fill. One permitting a more stable basis for the increasingly complex and larger scale technical procedures. Also, a

much larger population that would encourage diversification and specialization of the human operators and their sharing of ideas. And providing new sources of nutrients upon which technology could grow and prosper. All manner of ores, ceramics, pigments and resins. A location that would simultaneously benefit the human host.
One such place turned out to be the region served by the rivers Nile, Tigris and Euphrates. This area having the fertility to make farming a practical life-style for humans. Also, being sited upon the crossroads for many excursions to and from various parts of Eurasia, providing a wealth of exotic materials and ideas upon which all these infant technologies could feed. A significant augmentation of the dynamic network which identifies with the technosphere. With this far greater range of interactions, this wider sharing of imaginations, in conjunction with the additional time to invest in chemical and artistic experimentation and involvement in blue skies research (aka just fooling around with things), the scene was set for an equivalent of the Cambrian explosion in the evolution of ideas. The particular idea which would one day merge with many others to become the motorcycle was one. It embodied the notion of sitting astride a moving machine so as to avoid the necessity of walking or running. The machine in those times being that newly utilised biological tool, the horse. Which itself would soon inspire yet another tool, the saddle, and later, the stirrup. Which also made practical the use of this composite tool for humankind's favourite occupation, warfare. An idea thread which, in centuries to come would, after innumerable other interactions with the vast shared web of ideas, culminate in the battle-tank of today .

Starting with the very primitive idea represented by the winding of a "string" around a stick, an art probably used in a previous era for such practicalities as construction of shelter, the new diversified and highly networked human societies, as particularly exemplified by Sumeria, allowed the emergence of artisans. A new category of individuals, now freed from many of the more basic behaviours of humanity, such as foraging for food or killing other humans for revenge in the perceived service of their gods. A group to whom the input to, and exchange of, shared imagination was tremendously magnified. People who had time to twiddle. Twiddle with such things as pulling a string of some kind from a stick around which it had been wound. Thus making it rotate. The idea corresponding to this toy leading to the concept of an open bearing. In turn leading to a closed circular bearing. Already, we have a crucial precursor of much of modern technology, the axle. This idea no doubt pressed into service in the development of pottery, a technology also rapidly evolving at that time in conjunction with the activities of those that played with fire. The use of the axle inspiring the potter's wheel and an idea chain that would spread the axle notion, to eventually become ubiquitous in all manner of transport and machinery. However, we will follow another branch of that evolutionary tree, one that involved a very powerful lever. Returning to our rotating stick, it was just a matter of time before, by accident or on a whim, someone applied a sharp edge against the turning stick and thus the idea of the lathe emerged. Playing around with this newly evolved idea, spiral grooves would inevitably be formed. An idea that would diverge to an evolutionary branch that would further lead, among other things, to our favourite fasteners, the bolt and the screw. But we won't go there, either.

Having learned to produced a deep groove on a wide diameter piece of wood, somebody, maybe influenced by the idea of a bearing, perhaps chose to enhance this art-form by surrounding the groovy stick with a piece of wood having suitable protuberances such that it would whirl its way down. A toy from which the idea of a wine press would evolve. And from that idea came the crucial actualisation of the printing press, which by massive amplification of the sharing of imagination, would radically change our world.

Just as with the biological phase, some aspects of technological evolution are punctuated, either by contingency or by the necessity of waiting for an appropriate niche or for development of other parts of the network to provide input. Contingency is exemplified by the conflagration that destroyed the great library of Alexandria around 300 CE, in which, so it is believed, then also resided the significant collection of scrolls previously located at Pergamum. The final blow to what documentation still remained available in another important

Alexandrian repository, the Serapeum, was the destruction of much of that collection by religious zealots. The resulting damage to the technological network was profound. Many intricate idea complexes were completely lost. One such lost concept was that manifested in the antikythera mechanism, an analogue computing device:

*Fragment of Antikythera mechanism*

This contrivance, found in a shipwreck, and having more than thirty bronze gear-wheels, has been dated at around that same time and had a level of technical sophistication that would not re-emerge for more than ten centuries.

As for "waiting to catch up": Humankind had long harboured the idea of emulating the birds. Myths incorporating this idea have their origins exceedingly far back in the mists of antiquity. Their proposed embodiments include magic carpets, broomsticks and the saddling up of giant birds. By interaction of these with other ideas arising from the serendipitous emergence of the lateen sail, more practical implementations of the idea were notably suggested around 1500 by the Italian polymath, Leonardo da Vinci, a person exceptionally skilled in exporting his bountiful imagination to graphic images.

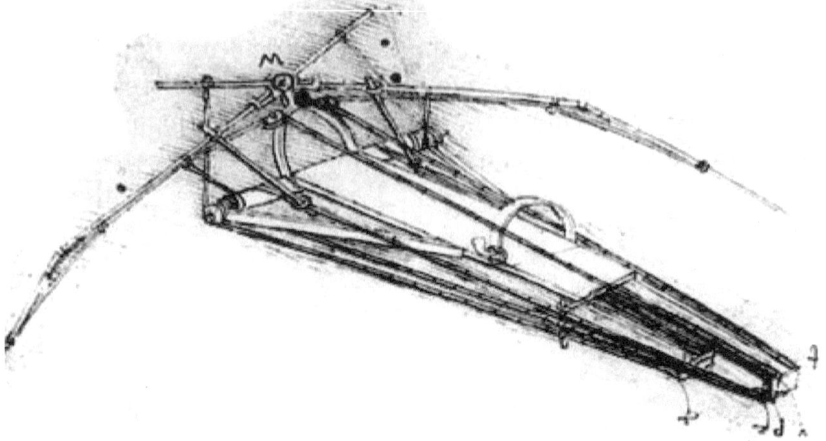

*Leonardo glider*

The advances in windmill construction triggered by this new sail idea, which employed aerodynamic lift, rather than drag led to the construction by George Cayley of a glider that first flew in 1804. Although many other functional gliders were built, the dream of useful sustained flight by a heavier than air contrivance was to wait another half century before the intersection with another very active idea complex, the internal combustion engine. The very rapid evolution of which was to produce a unit sufficient light and powerful to at last fully realise the ancient idea of powered flight.

*Wright Flyer*

(The idea of flight had a parallel developmental chain, of course, in lighter than air devices. the first powered flight, using a steam engine, having occurred in 1852. But that is another story!)

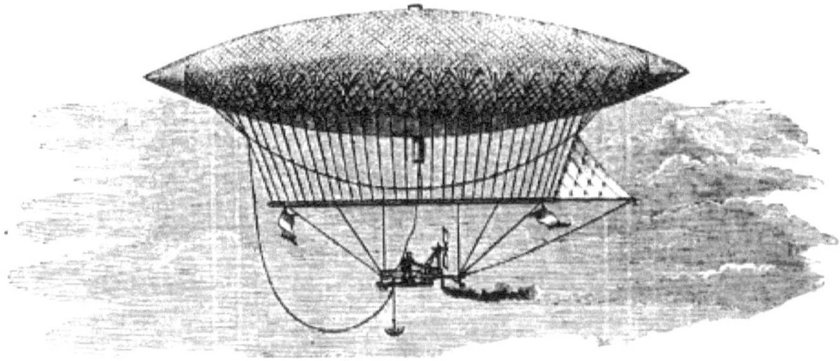

*Giffards steam powered dirigible*

Though certainly often intermittent on the large scale, both biological and technological evolution almost always occur by gradualistic processes that take place in tiny steps, even though this is not always immediately apparent. It is the rate of change, the rate at which the consecutive steps occur that varies. Also, as far as we can tell, there is no turning back in the development of either of these sub-networks or, for that matter the evolution of the universal network as a whole. In the words of Maurice Daumas:

> *"if we consider the history of technology at the scale of human history, not just of certain civilizations, we never detected a regressive evolution of technology"*

and that:

> *"since the beginning of humanity, technological progress continued steadily, almost flawless "*

The development of technology is utterly dependent on chemistry and geometry and its progression is tightly constrained by these. Of course, it also needs shared human imagination for its medium. But this medium itself is, at its roots also entirely defined by chemistry and geometry. That all these entities are "just right" for biology or for technology should really come as no surprise. Our science is based on what we observe. On "what is" rather than "what might be". The carbon atom is a telling instance of "what is" and its entire repertoire of unique properties is absolutely essential to biology. There are inevitably some who will shake their heads sagely, saying "Aah, that's for biology ***as we know it***!" But that is the ***only*** biology we know of, or that we can even start to fit within the framework of the big picture of consilience. If we admit of metaphysical biologies (or universes) we may as well admit the existence of the Flying Spaghetti Monster or go back to worshipping Thor, the god of thunder.

Keeping our feet firmly on the ground, while taking Richard Feynman's advice "to always keep the door just ajar for new evidence", we must, in any scientific context, admit that Not only carbon, but hydrogen, with its eponymous bond among its range of properties and also oxygen, are absolutely crucial for biology to eventuate. Without a shadow of doubt they are "just right". As, similarly, are nitrogen, phosphorous and many other components of the atomic assembly kit.

In the same way, when we come to the evolution of technology, which, stemming from biology and thus having the previously mentioned elements as prerequisites, has its own set of "just right" entities without which it could not have occurred. Perhaps the most clearly obvious is the prior existence of an immensely intricate biological network containing a subsidiary network with nodes very much along the lines of that of the imaginatory, language enabled, snoutless ape that is spread so wide on this planet today. But even within this prodigious and undeniably "just right" medium, technology cannot just spring up from thin air. We have looked at some of its requirements in earlier times, wood, fire and so forth. But when it reaches its later stages, there are certain newly required materials that are required to be readily available for it to make any further progress. The three of these most distinctly identifiable, the "holy trinity", are, in this case, transparent glass, copper and iron Without the "just right" properties uniquely exhibited by these substances we can be quite sure that the process would have ground to an abrupt halt That modern science and technology simply could not exist!

Those who have read "Unusual Perspectives" may well recognize the next section as essentially a somewhat modified and considerably extended recapitulation of the material previously presented therein:

So far we have been looking at the role of some key substances directly involved in supporting life, the structures of organisms, the life process itself, and establishment of, among many other species, our own. The further we come down this path, the more closely we look, crucial "just right" properties become ever more numerous and increasingly suggestive of a process having an inherent directionality corresponding to the action of our stochastic ratchet.

# Silicon

Silicon is one of the most abundant materials on our planet and, in combination with sodium, calcium, aluminium and other elements, comprises most of the rocky and sandy surfaces of our planet. Because of its similar structure to its far more versatile cousin, carbon, each four armed atom is able to form multiple links both to itself and to other kinds of atom. It is to minerals, what carbon is to biology and is the skeleton of most naturally-occurring non-living polymers. And it was with forms of these two kinds of widely found polymers, wood and stone, that the ancestor of humankind's first primitive technologies, tools and weapons, began.

A great deal of water was to flow under the bridge before the imagination of our species had developed to the point where the ability to conceive of and to practice the fabrication and use of significantly different materials had arisen. But when that time came, just the right stuff was, as usual and most fortuitously, right at hand.

No great inspiration on the part of any individual was required for this to occur. Our race had fire, and the blending of fire with both the emergent imagination and the ubiquitous silicon-based clays would, mostly in tiny steps, inevitably assure the development of the embryonic ceramic technologies to steadily progress, almost as if of their own volition.

One of the problems we humans encounter, when trying to figure stuff out, is that we have a very strong tendency to fall into the trap of taking things for granted. Oft-times losing the sense of wonder. We tend to regard clay as just some boring old muddy stuff with nothing particularly remarkable about it. And in a way I suppose that's right. But without that kind of muddy stuff that can be moulded by hand and gets very hard when you put it in a hot fire for a while and, specifically, without an abundance of polymer-forming silicon, the art of pottery would not have been born.

Most importantly, the smelting of some of our metals was initiated by, and depended upon, the existence of pots and almost certainly could have not eventuated without the silicon based clays. There are ceramic materials which don't contain silicon. But none of these would have been likely candidates for humankind to have been able to use so as to get its first foot-holds on the ladder of modern technology. So, again, we have an element tailored "just right" to form the minerals that have ensured that we would move forward. We will look at other important aspects of this very special element later.

# Copper

The upper crust of ancient Greece had an interesting party trick. They would take a ball of amber, a fossilized resin originally exuded from trees, and insert a spindle through the centre so that it could be set spinning. If the ball was then rubbed it was noticed that a spark was produced.. It was not until the eighteenth century, however that various individuals, who happened to have the financial wherewithal, the spare time, and the inclination to fool around with this bit of magic, began to try to make some sense of it all.

A little earlier, a certain Otto von Gueric had built a fancy version of the old Greek toy using a big ball of sulphur mounted on a mechanism that could cause it to rotate at high speed. Friction applied to the spinning ball would produce a very hefty spark. Tinkerings along these lines had served to revive considerable interest in the phenomenon. It was further noticed that these charged objects would also either attract or repel small shreds of paper. The ball could sometimes also be seen to glow.

Among those greatly intrigued by these magical behaviours was the alchemist, Isaac Newton, and he, among others, came to the conclusion that some kind of fluid, an "electric spirit" was strangely transferred by these manifestations. It was not Newton, however, but one of his disciples, the German Petrus van Musschenbrock, who most famously advanced the understanding of this "electrical fire" by finding a way of containing and storing it. He, among others, rather predictably came up with the idea of trying to somehow transfer it into glass bottles for storage. To this effect, attempts were made to "dissolve" the electricity in a jar of water. While playing around with this concept, the serendipitous discovery was made that no water was required as long as separate conductors lined both the inside and the outside of the glass jar. To fill the jar with the electric fluid you just hooked up the output from the spinning ball or disc to the inner conductor and the outer to the "earthy" side of the generator. There you had it – electric spirit in a bottle – a Leyden jar, to be carried away and used when wanted. This provoked a lot of interest and lots of people who happened to be in the know started trying all manner of things with their bottles of the mysterious invisible fluid which had been dubbed "electricity".

The French clergyman Jean-Antoine Nolett, among his other talents, appears to have had a flair for showmanship. He seems to have caused great amusement at the Versailles court of King Louis XV by emptying one such bottle of the "electric virtue" through a chain of 180 soldiers holding hands, causing them to leap into the air simultaneously. Enormously impressed, Louis ordered a repeat performance in Paris, this time with 700 monks. Perhaps inspired by this, the Italian anatomist and surgeon Luigi Galvani tried passing current from his bottle through a frog's leg, resulting in contraction of the muscle. On another occasion, however, he was surprised to notice a somewhat similar twitch while dissecting a frog's leg when no external electricity was around and jumped to the conclusion that the electricity was somehow produced by the dead frog itself. It was his side-kick, Count Volta, who figured out, to Galvani's disbelief and displeasure, that the electricity was, in actuality, produced by a reaction occurring between two dissimilar metals, in this case a brass hook supporting the leg having accidentally come into contact with the steel scalpel. Replacing the frog with saline impregnated fabric and sandwiching this between dissimilar metals he produced the first electric cell, and by stacking such cells together end to end, the first electric batteries. This was all very intriguing, of great interest to theorists and wonderful new frolics for experimenters. Also, these new batteries provided a much more versatile and controllable source of electricity. A very useful feature in the laboratory, where hitherto impossible feats such as the separation of metallic potassium from its salts could now be realized. But all might have soon stalled. A key component was required to permit the kind of development that would lead to the immensely useful large-scale electrical systems by which most of today's technologies are driven. The prerequisite for any such development was a substance with the following properties:

- It must be a very good conductor of electricity
- It must be capable of being drawn into very fine wire
- It must be resistant to corrosion
- It must be a good conductor of heat so that such components as (the, then unimagined,) motor windings and transformers that were eventually to be created would not cook.
- It must be easy to join, and the joints must be both reliable and conductive.
- It must be abundant, because, in the course of time, vast quantities would eventually be required.
- It must be fairly easy to extract from its ores so that costs would not be prohibitive.

A tall order indeed! But yet again the one and only material that would meet these demanding requirements was right at hand. To trace the inevitable origin and progress of this "heaven-sent" substance we need to go back 6000 or so years

There are few metals to be found in their native elemental form on, or near, the Earth's surface. Gold, platinum and silver can be found in this state, along with some iron that has arrived in meteorites. Copper, too, sometimes occurs in elemental form so was known in antiquity. Thus plenty of people were around who had acquired the know-how for shaping that metal by hammering and subsequently re-heating in order to remove the embrittlement caused by the hammering. Some seven thousand years before the current era the territory called Sumeria occupied the land that laid between the Tigris and Euphrates rivers. Soils swept down these rivers by floods originated in the mountains of Turkey and provided exceptionally fertile soil. With imaginations available to implement irrigation systems, agriculture blossomed, and, by the standards of the day, the living was easy. Surplus food was available to be traded for other commodities and, being nicely located on the established migration paths from Africa through to Asia, India and Europe, easy access was available to trade routes by sea, by river and by land .

It was *the* place to be. Real estate with the three big "L"s. Location, Location and Location. People came to settle from far and wide, forming a great civilization and a convergence of skills and ideas and technologies. And, as this was a land of plenty, at least some of the population would have had the spare time to play around with metallurgical methods, to interact with travellers, and to more easily obtain unusual materials from afar. This intense activity and cross-fertilization of Imaginations indubitably inspired many novel processes and devices. Wars, of course, there were many. But smaller proportions of the population would be pressed into soldiery and direct food seeking than was the case with the previous nomadic hunter-gatherer lifestyles. Furthermore, the various, now far more wealthy and powerful rulers of cities had, no doubt, become well aware of the value of the new artisan classes not only to produce trinkets and useful domestic items but also to come up with ever better bits of kit for killing people. Thus the more creative and inventive types acquired some degree of respect and immunity, giving them a reasonable chance of emerging unscathed from the bursts of conflict and relief from menial toil.

And no doubt, among this new class of artisans, a technology of particularly high interest was fire. Established methods, fanning, blowing with pipes, and funnelling the breeze in some way, would all be brought to bear, but somewhere in the course of the fiddling and twiddling, the operation of the stochastic ratchet, by some contingent sequence of serendipitous steps eventually led to the recombination of ideas that became expressed in the form of the bellows.

An even more important trick discovered in the quest for the hottest fire on the block was the preparation and use of charcoal. Charcoal is made by heating wood without letting it catch fire. In that way all the water and other volatile rubbish is driven off, leaving behind pretty well close to pure carbon. Now, accounts differ on exactly how hot you can get a wood fire, but, even with the best set-up and employing bellows, it probably maxes out around 700°C. That's enough to melt lead and also to release it from its silvery coloured ore, galena, or for annealing hammered copper. But that was about it for the guy who liked fooling around with metals until these exciting new high tech charcoal fires came along. Then, while pottering around with pottery associates, temperatures of perhaps 1000°C could be achieved.

For the snoutless ape, endowed with a keen curiosity born of language and its conjugate, hypertrophied imagination, there was great fun to be had in, for instance, putting lumps of different kinds of rocks, or copper, or silver into ceramic vessels to see if they melted or what. Sure enough some of the metals did, and from the rocks sometimes appeared new metal. Here, yet again, nature's discreet machinery was behind the scenes, dispensing liberal helpings of serendipity. For even at this higher temperature, another component was required to successfully release the metals from the ores. And what should that essential component turn out to be but our dear old friend, carbon.

What a stroke of luck! All that charcoal used for the hot fire just happened to be floating around to accidentally (at first) be spilled in the pots! The presence of carbon in the confinement of a pot is an excellent scenario for the production of carbon monoxide, which is exactly the sort of thing you need to reduce heated ores so as to free the metals. The serendipity doesn't end there, but as we now have the beginnings of a means of producing an abundant supply of copper from its ores and also being within reach of its melting point which, for the pure metal, is 1083°C, we shall shortly move again forward in time.

First, though, we'll just tell yet one more little serendipitous tale. The much smaller amounts of native metallic copper which had been used for a long time previously was quite good for making ornaments and pots. But it wasn't any great shakes for making very useful daggers or knives, as the metal is too soft to hold a good sharp edge. And probably the best way of coming up with the equivalent of a good research grant in those days was to be able to go up to whatever Godfather happened to be nominally in charge of your city with some exciting new device for killing people. Copper had serious limitations for use in weaponry. But all was not lost! The metal workers of Sumer were fortunate in having local access to both arsenic and tin. Either of which if dissolved in copper give the alloy, bronze.

Bronzes have quite different properties from pure copper. For starters, the presence of tin or arsenic lowers the melting point to a level quite easily attainable by a forced air fed charcoal furnace and makes casting an option. Furthermore a blade made of bronze can hold quite a good cutting edge and is thus very handy for slicing up your fellow man. Which probably pleased the Princes no end. While arsenic is often found in conjunction with copper, this is unusual for tin. The extreme toxicity of arsenic no doubt discouraged that approach, but those lucky Sumers happened to have tin right on their doorstep. At Susa, one of those very rare places where tin and copper are found together. So the tin-arsed Sumers, who by virtue of their excellent location on the trade crossroads would, without doubt, have hit upon the tin-copper bronzes eventually, were thus able to steal a march on the artisans of those in another significant hot-bed of technological development in those early times, the Majiayaore region in the fertile upper Yellow River valley of China. An extremely productive culture whose early technologies also included the use of the blast furnace for producing a range of steels as well as paper-making, printing, the seed-drill, the plough-share, explosives, drilling for oil and many other processes which presaged the eventual emergence of similar skills and devices many centuries later in the very extensive interactive networks of Western cultures. But enough! We have lingered in the ancient times all too long. Let us travel forward in time again to eighteenth century Europe.

Up till the time of Michael Faraday, all of the twiddlings with things electrical, based on electrostatic generators and electro-chemical batteries, could have been achieved with any of a variety of metals. Wires were proving to be the most convenient way of piping the currents around but, even so, there was quite a good choice among the then available metals. Gold, silver, and iron, as well as copper, could be quite easily drawn into wires and were all used. But an entirely new way of producing electricity was soon to arise, and the enormous extent to which it would come to be used would impose constraints such that copper alone, among these metals, could meet the prodigious future requirements. And, of course, as we have seen, much of the essential groundwork for copper processing had, very conveniently, long been in place.

The entirely novel way of creating electricity resulted from a chance observation in 1820 by Hans Christian Oersted that, whenever he switched an electric circuit on, the lodestone in a nearby magnetic compass would suddenly swing round. Speculations regarding some connection between electricity and magnetism had been previously advanced by various philosophers, but here was direct evidence. Among those intrigued by this phenomenon was Michael Faraday. An assiduous and enormously imaginative experimenter, he firmly established that a changing electric current could cause movement of a magnet and that, conversely, movement of a magnet placed near a conducting wire could cause a changing current to flow through that wire. Thus, electricity and magnetism were two sides of the same coin! The effects, while reproducible, were too small to be useful using single straight wires. Faraday found, however, that the effect could be magnified by the simple expedient of winding the wire into a coil. The more turns you had, the correspondingly greater the effect. By having a very large number of turns and rotating a suitably oriented magnet you could generate very substantial amounts of useful electricity. All this without recourse to batteries or static devices.

There was one little problem. Because the magnet was rotating, first one pole, then the other would be inducing the current. So, the current, rather than flowing continuously in one direction, instead surged back and forth. But it was no great problem to find a way round this and, in any case this surging, alternating current (AC) eventually turned out to be of greater utility for far more many purposes than the steady flow of direct current (DC). Using the reverse principle, Faraday devised the electric motor as well. And by means of the new AC he was able create the device known as the transformer which, as well as being essential for the efficient transmission of power to our homes and industries, has been an indispensable component of nearly all the electronic devices underlying the development of most of our modern technologies.

Sure, certain applications of the transformer have now been superseded in today's electronics, but we could never have reached that present state of the art without the transformer to first carry us to that stage of development. And for all this, copper is the only real option. Silver and gold are not available in sufficient quantities. Iron corrodes quite easily, is nowhere such a good conductor of electricity or heat, and is not particularly easy to join. Metallic aluminium, in Faraday's time, was still in its birth throes. In fact it required the development of electricity to produce it, either directly, from the electrolysis of Bauxite as practised today, or indirectly, by reduction of the ore with potassium (this element itself having been produced by means of electricity). Aluminium is ductile and a quite good conductor but is nearly not as good in either respect as copper. It is much harder to join reliably and is more reactive than copper, so has poorer corrosion resistance. It is, however, these days, very abundant. Aluminium has occasionally been used for the windings in some small electric motors. Because of its much lower density it is widely used for overhead high tension power lines, where weight is a major consideration. Also, for a while, in the USA, for house wiring, but not without some serious problems. So it might naively seem that aluminium wire based electrical and electronic technologies might have somehow precariously stumbled along. But this is actually not the case, because without copper we would have no aluminium worth speaking of, anyway. We are thus forced to the conclusion that copper is the only really the only option for use in the windings of the many billions of motors, solenoids, transformers, generators, loudspeakers, relays, printed circuit boards, cathode ray tube yokes, and the countless other such devices so characteristic of our world today. Not to mention such applications as general electric power distribution, and telephone lines.

Yet again, with copper we have what seems to be exactly the right, virtually custom made, material and that huge quantities just happened to be right at hand to enable the continuance of the observed very long-standing evolutionary process. The snowballing growth of the network of which technology is now the salient phase.

# Aluminium

This is the most abundant metal in the earth's crust. It is a major component of those clays that seem to have taken us by the hand and gently led us into the development of ceramic technologies. It also occurs in a wide variety of other minerals. But until the nineteenth century the raw metal had never been seen. At the 1850 Paris exhibition a single bar of the metal was proudly displayed and with fanfare, billed as "the new precious metal". Indeed, at that time its price was higher than that of gold or platinum. The harnessing of electricity has now allowed us to prise it out from its ores, and as the efficiency of this process steadily increased, the price dropped very dramatically. Because of its low density, and the reasonably low effective reactivity that results from the protective oxide layer which quickly forms on the surface, it has a wide range of applications.

The special application for which its various alloys are most uniquely suited is aviation and for space exploration. While it is certainly a rather fortunate coincidence that it was to arise on the scene in quantity just about the right time for that use, it cannot quite be so convincingly held that the lack of the metal would represent a bar to the development of technology as a whole. Even its widespread use in mirrors, particularly those in telescopes and other scientific equipment could just possibly have been worked-round, albeit more far more clumsily, by the use of, for instance, silver. It certainly has been a great enhancer of recent technology but perhaps cannot quite so easily be claimed to be a critical item. Its role in combination with silicon and oxygen as clays in earlier times is rather better placed to hold that distinction.

# Iron

An account of the host of subtle "coincidences" that permit the modification of iron into a whole raft of alloys having unique properties could fill a whole book in itself. We shall review here just a few of its other more salient qualities :

You may remember, during our visit to ancient Sumeria, that the coolest cats had forced air fed charcoal furnaces that could, under ideal conditions, perhaps be tweaked up towards 1100°C or so. It is likely that they were particularly attracted to the various iron ores, some of which are quite brightly coloured and were no doubt already in use as pigments in those times. Also by the sulphide ore "fools gold", iron pyrites, the appearance of which deceptively hinted that the desirable gold, with which they were well familiar, might lie within.

But although the iron might well be released from ores by carbon monoxide at the temperatures attainable in their furnaces, the metal so produced could not liquefy. The melting point of iron is way up around 1500°C. With a few tricks up their sleeves, derived from their smelting of less difficult metals, however, particularly the "bung in some charcoal" ploy, it would be merely a matter of time before somebody tried, say, mixing some ground up iron oxide pigment in a ceramic bowl with a nice lot of charcoal powder and heating hell out of it. The satisfaction of winding up with a lump of metal being perhaps marred by the realization that it was full of bits of rubbish consisting of the left over siliceous and other impurities from the ore and, in that state, not very useful. Perhaps a particularly frustrated smelter became so incensed that in a paroxysm of "iron rage" he threw the glowing porous mass from the fire onto a rock and thrashed the living daylights out of it in retribution. Or perhaps the discovery was more prosaic. But, however it happened, it was found that repeatedly hammering and re-heating the porous mass was quite effective in the removal of the junk impurities from the metal.
The crystal structure of iron formed in this way is such that the metal is quite malleable and, especially when heated, can be shaped fairly easily. It is not particularly hard, though, and being much more laborious to smelt, did not supplant bronze for a considerable length of time. It is ironic (!) that, because of another handy coincidence, the production of a more versatile iron preparation may have been just within their reach. For, although pure iron melts around 1500°C, it will, at high temperatures, start to absorb another material that will lower the melting point to just over 1100°C. The resulting alloy, when solidified, is much harder and more useful. But did the metal workers have this special substance that was "just right"? Bet your sweet life they did! Just good old charcoal carbon again! So the Sumers seem to have narrowly missed out on this one, and the honour of discovery was to be attributed to a Chinese community, who managed to get that little extra temperature out of their furnaces and to get in first with steel. For, as usual, this next advance in iron technology was all in place, waiting for its inevitable serendipitous eventual discovery.

The ferrous metals have become such an integral and all-pervasive part of our civilizations that we would find it hard to conceive of a world without them. In fact they are so common-place that, as is our habit, we just take them for granted. And if asked the question "what would we have done without iron" a common glib response might be that "We would just have used something else". But any such naive assumption is, the more closely you look at it, completely without foundation. For, again, there ***is*** nothing else that could play that role in allowing the development of technology that has occurred in the last two millennia. Sure, we can name the odd gadget here and there which allows substitution of another material But in the big picture iron, and its many daughter alloys, are an indispensable feature. To try to list them all would be both time-consuming and trite. Let's instead just look a few of the more critical features along the various branches of endeavour from which today's civilization has evolved.

**Magnetism:** The discovery of this phenomenon depended upon the existence of the lodestone. This is a form of the iron ore, haematite. Not all haematite is magnetic, the lodestone variety, first noticed by the Chinese, is believed to be given that property as result of lightning strikes. Planned ocean journeys and cartography would have been greatly hindered, if not made impossible, by the absence of the magnetic compass. More importantly, work of the kind carried out by Michael Faraday would perhaps not have been initiated and most certainly could not have been completed without the magnetic material, iron, which is the only option for motor and generator armatures, transformers, and solenoids The use of electricity would have necessarily stalled at the stage of static generators and batteries. Sure there are a few other ferromagnetic materials, but the discovery, use and production of these have all depended on, for one thing, the ready availability of abundant electrical power. The equation is as simple as it is undeniable.

**No iron = No civilization based on electric power.**

Furthermore, for both intellectual and practical reasons, no significant development of electronics.

Not only would the development of electric power and electronics have been impossible without the "just right" properties and ready availability of iron. It is also of crucial importance to many other areas of technology. For example:

**Civil and mechanical engineering** : Without iron and its family of steels we would have no steam locomotives, no rails, no steel or steel reinforced bridges, no heavy earth machinery, no motor cars, no steel ships. Anything like our kind of civilization? Forget it! Or perhaps you think we could have used aluminium instead? Well, in view of its softness and other inadequacies that makes it unsuitable for the kind of machinery that requires the properties of steel, how exactly are you going to dig out the bauxite and transport it to the smelter? Once there, then how, lacking the magnetic qualities of iron, are you going to generate the huge amount of electric power required to melt the ore and to perform the necessary energy-intensive electrolysis? From a battery? For in the same way that water is seen to be a progenitor of most common liquids, iron and copper are progenitors of metallic aluminium.

**Medicine:** Iron is, of course a biologically important element and plays a special role there. But even at the cutting edge of medical science we have the surgeon's scalpel with its, almost invariably, steel blade. There is, actually, in this case, a very good alternative in the form of obsidian, a volcanic glass. Flakes of this material are actually sharper than the very best steel and, for this reason, are sometimes used in eye and other very delicate surgery. It is not so readily available, though, and is much more difficult to mount in a handle. But obsidian is certainly not a practical replacement for the steel knife used in a microtome, which is a gadget for cutting very thin slices of tissue that are suitable for viewing under a microscope. This device is important in diagnosis of disease as well as in research environments. A more recent and very powerful diagnostic tool is the MRI scanner which uses the interaction of powerful magnetic fields with the nuclei of certain atoms to generate three-dimensional images of internal structures. The magnets in this case are not made of iron. But that's not the point, for without the lodestone and subsequently the iron magnets, not only would we have no power outlet to plug the scanner into but we would also never had acquired the know-how necessary to create such a machine. Nor have been able to acquire and fabricate the very special materials from which some of its key components are made.

**Commerce.** Transportation is one of the keys to commercial activities and these were thus greatly enhanced by the use, in earlier times, of horseshoes and iron wheel rims. Today's commerce is highly dependent on electronic data processing. One link in the chain of development of the computers which now carry out, for instance, electronic funds transfer, was the magnetic data storage tape. This, a version of the recording tape, once an important part of the radio, television and music scenes, helped bridge the gap between the punched cards or paper tapes, which were becoming far too slow, and the more recent floppy and hard drive storage devices. All these depend upon magnetizable iron oxide grains for their basic operation. It is possible that less efficient work-arounds might have been found for either of these two very different facilitators of commerce. However, particularly in the case of magnetic storage devices used for computers, it may well be that the wide-spread use of these might not have even taken off. In which case the subsequent development of the optical and solid state electronic storage devices we have today might not have eventuated. For these evolutions are not as simple as we often naively assume. There are so many intricately convoluted interactions and ramifications. Wheels within wheels!

**Agriculture.** There is a considerable cost in the development of new technologies, although they are investments that nearly always pay off, often quite handsomely. It is not only armies that march on their stomachs. The primary produce of farming is an important component of any economy and the back-bone of some. The iron plough-share alone revolutionized agriculture. Slicing and separating the soil in a manner far superior to that achieved by the pointed wooden post previously used, and easier to push or pull. The iron and steels, so ideally suited for this and related tasks, gave rise to a host of agricultural instruments, culminating in the steam traction engines and then tractors that would pull them more efficiently than the horse. The combined harvester, the fruit-picking machines, so it goes on. No real alternatives for iron here. We have only scratched the surface but can already see an iron-clad case for the absolute necessity of iron for the development of a technology anything like that which we have today. And the more closely we look, the more that iron-clad case hardens.

# Glass

We are back again in the company of our old friend silicon. But the properties of this particular group of its polymers is so very strikingly suggestive of purpose that it could be considered the jewel in the crown of apparent teleology. Progress with all the materials we have looked at so far have all, of necessity, been closely associated with a concurrent growth in science. To use things, and to find better or new ways to use them, you must inevitably learn about them, by trying things out and making observations. Each of these newly made observations, unless the witness is unnaturally secretive, adds to the current fund of shared imagination. And the increase in collective imagination, in either natural language or that of mathematics, leads to even more observations, discoveries and new capabilities. Here we find again yet another process which is self-perpetuating and usually also self-accelerating. The kind of material we call glass, however, goes much further than that. For if it could be shown to have, or is ascribed, a" purpose", then the very essence of that "purpose" would seem to be the propagation and enhancement of shared imagination.

Indeed, in many instances it does that directly, by expanding the input range, the sensitivity and selectivity of our native senses. As always, the materials for its manufacture are very easily accessible.. The required silicon pre-polymer pellets that have been ground, in the course of many aeons, by the action of sea and wind to a most convenient size for reaction. We call them grains of sand.

The other essential components are just where you want them too. They are the main components of the ashes of a camp fire - mostly potassium and sodium oxides and carbonates with maybe a reasonable amount of calcium compounds as well, if some sea-shells or bones have been chucked in the fire. Not every time you make a camp-fire do you wind up with glass, of course. But conditions vary widely, and we humans have a strong proclivity for fooling around with things. With such propitious circumstances it was inevitable that glass would be discovered, not once, but many, many times since our species has been playing with fire. We seem to be attracted to shiny or sparkly things so it was likely the little blobs sometimes created were kept and treasured. Those people, no doubt, attributed magical properties to this substance. How right they were! Gradually, set-ups would be found that would greatly enhance the production of these objects. Although it is sometimes described in terms of a supercooled liquid, glass is much more usefully viewed as a thermoplastic polymer. One that can vary considerably in composition. Unlike a pure metal it has no specific melting point. As with the carbon based thermoplastics, say polythene or Perspex, it softens and turns into a viscous liquid rather gradually. The temperatures at which these structural changes take place depend on the length of the polymer chains and the extent to which they attract or impede each other. Some glasses with high sodium levels can become liquid as low as 500°C. These often have poor water resistance and some, in fact, can be dissolved. As higher furnace temperatures were achieved the versatility of the alkaline silicate polymers increased.

Around the time of the Sumers, the potential for glass-work and development was already high and some fine decorative items were produced. Significant later advances then occurred in Phoenicia and subsequently in Venice, but development has moved forward fairly steadily into modern history. An important, more recent, discovery was that the incorporation of borax in the composition greatly improves the resistance of the glass to thermal and mechanical shock, an effect of great significance for some of the key applications we will be considering. Again a "just right" coincidence because the bonds formed by boron have the unique ability to, in a sense, "walk" along two polymer chains bridged by this element and thus relieve stresses wherever they are induced. So far we have indicated and underlined the absolute inevitability of the "discovery" and the subsequent development of glass of good quality.

Perhaps it should also pointed out that, like the many other thermoplastics that now we have today, it can be formed and shaped fairly easily in many ways. These include casting, rolling, drawing, pressing, free blowing, blow-moulding and compression moulding. But the real mission of this section is to show how this material has, in marvellous ways, vastly expanded our scientific knowledge of our universe. Furthermore, that in most cases this knowledge was not accessible by any other means. Glass has been very much our window to the world. Here are some reasons why:

- **The window:** This simple measure provided early scholars and researchers with protection of documents and equipment from the weather while still admitting daylight.

- **The oil lamp chimney:** This extended the working hours of the scholar and researcher.

- **Spectacles:** gave further significant visual advantage to many such workers.

- **The mirror:** While mirrors can be made from metals, glass is the material of choice. The special mirrors used in sophisticated instrumentation today still invariably employ glass matrices. Even those very special mirrors called diffraction gratings.

- **Watch, clock and instrument glasses:** Provided protection for delicate mechanisms.

- **The telescope:** Provided unimagined knowledge of worlds beyond our planet. The refracting telescope was largely superseded by the reflecting telescope, but the only practical substance to use for the dish of the reflector is glass. This is still true of the largest reflecting telescopes in major observatories. Not to mention the Hubble. The telescope has, perhaps more than anything else, revolutionized our way of considering ourselves in relationship to the universe.

- **The microscope:** The extent of the impact of this instrument on the advance of science and technology is incalculable. It has precipitated such enormous strides in so many diverse disciplines that I could not begin to do it justice. I won't even have the temerity to try!

- **The prism:**The absence of glass to permit examination of the resolved spectrum would have prevented, or at least very seriously impeded, the understanding of the theory of light. Without glass the theory of optics would probably still be unknown. Water ice and transparent minerals such as diamond and feldspar were not of a suitable form for appropriate studies to have been initiated. Materials such as polymethyl methacrylate are available today and now prisms and lenses can be formed from these. But, as with other aspects of glass's applications, it is almost certain that, without the prior understanding of such fundamental physical principles, the very development of modern plastics would not have been possible.

- **The spectroscope.** This instrument, too, has played a very important part in establishing modern physics and chemistry.

- **Laboratory glassware:** Having spent a fair proportion of my working life in chemistry laboratories, it is very clear to me that glass vessels and associated equipment have been absolutely essential for any appreciable progress in this discipline. Glass of the kind used for these purposes is remarkably unreactive, and because of this, almost all liquid compounds can be stored in it and all manner of reactions, including those requiring heat, can be carried out. Yet, because of its transparency, the progress of reactions can be readily monitored and appropriate observations made. Measuring equipment such as burettes, pipettes and volumetric flasks are, again because of the excellent transparency, inertness and dimensional stability, very effective. Distillations can be conducted using flasks, condensers and associated components, and glass is often the only option. Even now it is possible to substitute other plastics for only very few of the purposes fulfilled by glass. For many years the only practical and versatile instrument for measuring pH has used a probe with a glass membrane for that purpose. The list goes on and on. It still applies even today, when a lot of work in the chemistry laboratory is carried out using instrumental methods. The instruments themselves usually having crucial glass parts adorning their innards. In earlier times "wet" methods were used to a much greater extent and such glassware was absolutely essential. Even the humble test-tube was probably critical in acquiring much of our present knowledge of chemistry (and biology, for that matter). For how can you detect a colour change or a precipitate easily in an opaque vessel? Well perhaps, much less conveniently, by using something like a white porcelain evaporating basin. Maybe! But what is the outer coating of porcelain. You've got it - glass! Laboratory glass is attacked to any appreciable extent by only two common reagents, hydrofluoric acid and strongly alkaline solutions. For these, materials such as wax or polypropylene are better, but otherwise glass wins all the way. Furthermore it has the advantage of a very high softening point. Yet not so high as to preclude fabrication of quite complicated apparatus with very simple tools. It takes skill, but an awful lot can be achieved with just a blow-torch flame and a bit of puff.

- **The Electric Lamp:** This, of course, by providing abundant illumination indoors and at night, has facilitated all manner of activities, including some that have been significant in speeding technological advance and, concurrently, the field of shared imagination which is science. But there are many other instances where such advance might have been prohibited by the lack of this device. Harking back to the chemistry laboratory we must always bear in mind that such instruments as spectrophotometers, have been absolutely essential for getting an understanding of many natural phenomena.. These come in various forms, all with quite different applications, and an essential requirement for most is an electric lamp, either of the filament or discharge type. A glass or quartz (pure silica glass) envelope is, for all incandescent electric lamps, the only option. Indeed it was the only option for all practical electric lamps, including those which use a plasma, such as the neon, the xenon and the common fluorescent. That is, of corse, until advent of the light emitting diode and the laser, either of which, in the absence of glass, would exist. Simply because the science and technology required for their development, the great bodies of both chemistry and physics that allowed of their theoretical basis and of the devices required for their fabrication, was absolutely glass-dependent.

- **The Thermionic Valve.** Although some of the earliest radio receivers used a semiconductor, most commonly a crystal of galena, touched by a wire "cats whisker" the thermionic valve was a key component in the meteoric growth of electronics in the first half of the twentieth century and the technological bonanza that went with it. Could the leap from crystal and "cats whisker" have been made independently? No, because, for exactly the same reason given above, the novel semiconductor technology first developed around the nineteen-fifties depended upon (among other things. such as the microscope) advances in chemistry and, as pointed out previously, in the absence of glass, chemistry would, without any shadow of doubt, still be stuck right back in the stumps. Furthermore, the invention of the fabricated semiconductor device required a whole huge bunch of imagination, especially of the kind provided by, to give just one example, the synergy between chemistry and the many applications of the thermionic valve.

- **The Cathode Ray Tube:** OK, I know they're fast becoming extinct now, but the CRT was yet another crucial bridge to the point where we could the create plasma, LCD and LED displays which have replaced it.. The CRT gave us the oscilloscope, the television and later, computer monitors. But the military purpose which spurred its development was to give us new windows on the world. We could now tell how far away things were that were out of sight. Radar was one of the first applications. It is not only used for warfare and for tracking aircraft and checking vehicles speeds. It has had extensive and direct uses in the expansion of our scientific knowledge. Improving our understanding of the weather, for instance, and by means of the sonar variation on the same theme, mapping the sea-bed. But that was only the beginning, because now we could observe phenomena that were otherwise completely invisible to us. The shape of a radio wave, or of an AC current flowing through a conductor! An actual sound modulated radio signal right there on the screen! A wave of depolarization travelling along a nerve fibre! The poor bulky old CRT, of necessity glass, has run its course, played its temporal role in the expansion of imagination/language duality and passes the torch to our excellent new, ever more compact displays for which, ultimately, glass has been an essential and indispensable precursor.

- **The Cloud Chamber:** Devised by Charles Wilson in 1911 for tracking the paths of sub-atomic entities was crucial to the development of particle physics. Glass was the only option for containment at that time, and, even when the bubble chamber took over in 1955, was still the material of choice.

- **The vacuum frame:** – Not the very different device used for vacuum forming of thermoplastics, but a little-known complex bit of glassware with which I once spent many happy times while researching adsorption isotherms on hydroxy-apatites and other substrates. Perhaps not a big player in the overall march of science, but one for which there was no option other than glass. This material is thus found to have played its evolutionary part in all manner of obscure scientific endeavours, in set-ups which you and I have probably never encountered.

- **Gravimetry** has played an essential part in deriving the science of chemistry. At first glance, at least this area of experimentation may seem to have been immune from the requirement for glass. But for any really useful degree of precision the analytical balance was required. And, even here we find that the only practical option for providing visibility within a draught-free enclosure was, yet again, glass. Today, of course, there are many ways of getting round this. There are plenty of transparent plastics which can, and do, serve that purpose very well. And, because we now have electronic read-outs, weighing systems can be devised that have no requirement for transparency. But we must bear in mind that neither the synthesis of plastics nor electronic readouts would be available to us without the sciences and technologies that have glass to thank for their very existence.

There are, of course, innumerable other aspects of the use of this evidently "heaven-sent" material. The marriage of glass with another special element, the metal mercury, has been of tremendous help in learning to understand the principles of manometry and the nature of pressure and the vacuum, as well as being a most valuable tool for such practical measurements as these to be carried out in many other fields. Also for the quite easy measurement of temperature. These are possibly supportive roles rather than prerequisites for development because there have been other, admittedly far less convenient, ways to make some of these particular measurements. The possibility of a critical role for the glass/mercury coalition should not be completely ruled out though, because the combination has been used for other purposes and their earlier widespread use, even in my time, for the chemical analysis of gases certainly springs to mind. Motor transport, too, has been highly dependent on glass for windscreens, lamps and mirrors and this clearly will have had a very significant impact on the development of our technologies.

## Calcium

If the alkaline vents scenario for the transition of chemistry and geometry to the biological phase is valid, then the element calcium, in the crystalline form of its sulphate which we call anhydrite, was the predominant component of the cellular womb from which it would be born. Far later in the planet's history, sea creatures evolved for whom defence against predation was provided by shells formed from calcium combined with those very clearly "just right" elements, carbon and oxygen, playing yet another of their essential rôles in the ever-growing increase in intricacy of the biological network. As the ages passed, the innumerable calcium carbonate shells became compressed into the vast beds of limestone, chalk and marble that are now found in great abundance all over the world. Calcium, even later in the evolutionary story, with the emergence of the vertebrates, found another very important part to play. This time with another of our old "just right" friends, phosphorous, in the hydroxy-apatite structure which gives our bony endoskeltons their rigidity and their low solubility. The machinery of the evolutionary network, while fulfilling all these tasks for the advancement of its biological phase, had quietly also been setting the calciferous scene for the eventual dawn of technology. For which the calcium carbonate shell residues would provide not only such useful but not irreplaceable items as black-board chalk which, for a short while, found a very recognisable application in the transmission of imagination, but also a class of substances that, especially in conjunction with another old friend, iron, would become quite crucial for the continued efficient development of the technosphere.

Just as with the clays, and for that matter, all the other "just right" materials that have made the advancement of technology possible, we have this overwhelming tendency to simply take concrete for granted. To blithely just assume that, "of course it has always been there." Which, in common with the other key substances, is largely true. And this is really a most important aspect of the contention that I am pursuing in this book. The strikingly conspicuous availability of these materials that, in close conjunction with the human lineage's language-enhanced imagination, and particularly its propensity for playing with fire, ensured that their serendipitous discovery was statistically guaranteed. Limestone and silicaceous rocks are found in almost every part of the world. An indication of just how long they have been around is underlined by the discovery of almost "ready to go" concrete as far back as twelve million years. Not it this case, of course, by pre-human intervention but by the burning of adjacent beds of oil-shale and limestone. The burning of limestone producing calcium oxide, which is the basis of the cements which bind together the usually silicaceous particles which constitute the other main component of concrete. Just as with clay pottery and the glasses, the very easy discovery and use of the concrete family by the hominid fire-freak was truly inevitable. Indeed, in a sense, we can even go so far as to say pre-determined. There may be some who would say that concrete has little to do with the evolution of technology. They would be very, very wrong! The employment of concrete for various purposes goes back at least as far as the Sumerian settlements, was used by the ancient Greeks and, like glass, markedly started to get its mojo on under the determined auspices of those clever, if rather belligerent engineers, the Romans.

According to much of the myth-based tradition that still raises its uninformed head, the "invention" of the wheel heralded the dawn of technology. As discussed previously, the notion of the axle was actually the hub of that particular technospheric sub-network. The wheel, as used for transport, was not, in itself, a very useful item. The simple roller being more suitable in those times. To appreciate why we have to return to the mantra that "nothing evolves in a vacuum". Evolution is a network function. For the transport application of the axle to be used to any significant extent, to provide any advantage over the camel or the mule, roads were first required. Roads that were fairly flat. The Romans were particularly adept in this area, initially using stones to impart durability. But the seriously disruptive technology which tolled the death-knell for the use of the laden beast of burden for transport was the synergy of axle and the cemented stone or the concrete road.

The "just right" basic material, hugely abundant, effectively self-discovering in the potentiating presence of human imagination, language and tool-skills, has transformed practically every facet of our civilisations. The direct impact upon technology extends well beyond the facilitation of wheeled transport, of course. Even in antiquity it was often the material of choice in large buildings. Today it is ubiquitous and, particularly when it is used in conjunction with steel, uniquely makes easily possible the construction of tunnels, bridges, skyscrapers, airports, canals, and levees. As well as virtually all other large-scale aspects of our modern infrastructures.

The steel reinforcement itself provides the tensile strength which concrete lacks and, importantly, the because of the extent to which both the iron based and calcium based materials can be tailored by composition, the coefficients of thermal expansion can be easily matched. Concrete is currently used at the rate of over twenty billion tons a year. Unlike glass, its involvement with science is rather less immediately obvious. But most certainly it is there, discreetly blending, for the most part, in the background. For a little careful reflection leads us to wonder if, say, space shuttle missions could have been deployed without its use as a pad? Or if its non-availability would have precluded the construction of the large hadron collider at CERN? Or what else could have been used to provide a foundation for the 200" glass mirror at the mount Palomar observatory? Hmmm...

Calcium, ultimately because of its particular atomic structure, its shape, its specific reactions with carbon and oxygen, and the reactions and water solubilities of the resulting compounds, is certainly "just right" to enable this irreplaceable behemoth of technology, concrete.

# Silica

The kind of glass known as fused quartz, unlike the silicates, contains only silicon and oxygen. Because it has a much higher softening temperature it is far more costly to make and to form. But the extra cost can be justified in some applications owing to its very special properties. Unlike the more common varieties of glass, it is transparent to ultraviolet light and hence is mandatory for such items as far ultraviolet lamps. It plays important roles in, for instance, UV spectrophotometry and photolithography. Perhaps one the most striking practical feature of glassware formed from the material is its capability to withstand thermal shock. Rather impressively, a rod of fused quartz that has been heated till it glows bright red can be plunged into cold water and remain intact. Common applications utilizing this particular attribute are the quartz-halogen lamp and sheaths for heater elements.

Fused silica is also very strong. Fine filaments and capillaries can be drawn and they are able to be flexed through quite tight radii. In this form, the exceptional properties of this high purity glass make it very much a major player, even now, in adding great impetus to the exponential expansion of science and technology. In the field of chemistry it still uniquely enables many studies. From the cantilevers of atomic force microscopes, the UV transparent cuvettes for spectro photometry, the capillary columns used for high resolution gas chromatography and capillary electrophoresis. And not only it is the basis of the new optical fibre communications revolution but it is, at the same time, opening up entirely new vistas of understanding in such fields as non-linear optics. The silicon on which glass is based, as well as being truly "just right "for both these and the many other facilitators of development we looked at earlier, has, in recent times, found yet another, entirely new, critical role in driving technology, and our shared imagination onward at an ever-increasing pace. That is the role it plays in the semiconductor devices that emerge from, most famously, the place that we call Silicon Valley!

## The DIY network kit

Entities with the unique sets of properties possessed by copper, silica and its compounds, as well as iron and its alloys, are seen to be absolute prerequisites for the very existence of the fast-evolving technosphere. A dynamic network that clearly exhibits an incredibly high degree of directionality, an exponentially increasing intricacy of both its chemistry and geometry, which are the underlying phenomena that find their ultimate practical expression in such things as electronics and mechanical systems.

It is very probable that, on sufficiently careful analysis, we could find many other elements or compounds which play an essential rather than a supportive rôle in the evolution of technology. Calcium, we see, is a strong contender. And it also may be the case that the sums of properties of certain elements may be similarly crucial. Exactly what other parts of the elemental construction kit are essential we are cannot be sure. Mercury is certainly a possible candidate. Another is tin, particularly in combination with lead or antimony for providing the robust and conductive joints required for copper wiring. and for printed circuit boards. Tin also having the "just right" properties for float-casting sheets of glass. Germanium was used in the early transistors but its non-availability would have hardly been likely to have seriously hindered the development of semiconductor devices. Similarly, uranium, although essential in having provided a warm habitable planet in the first case, might, in its absence, just conceivably not have directly scuppered the overall progress of the technology phase. Again, there are a number of elements, the transition metals, for instance, that have considerable overlap in some of their properties that gives some degree of redundancy. Some others might be considered as spin-off, their presence simply being inherent in the ordered sequence of atoms. Nevertheless, even of those whose function is not identifiably critical, most are found to play significant parts in the technological and biological ecosystems.

But certainly, as is the case for carbon, oxygen, hydrogen, nitrogen and phosphorous in the biological phase, the absence, or limited timely availability, of copper, iron and the glassy compounds of silicon would, without the slightest shadow of doubt, have been absolute show-stoppers.

# THE EVOLUTION OF MINERALS

## Phase 3 - Biology assisted

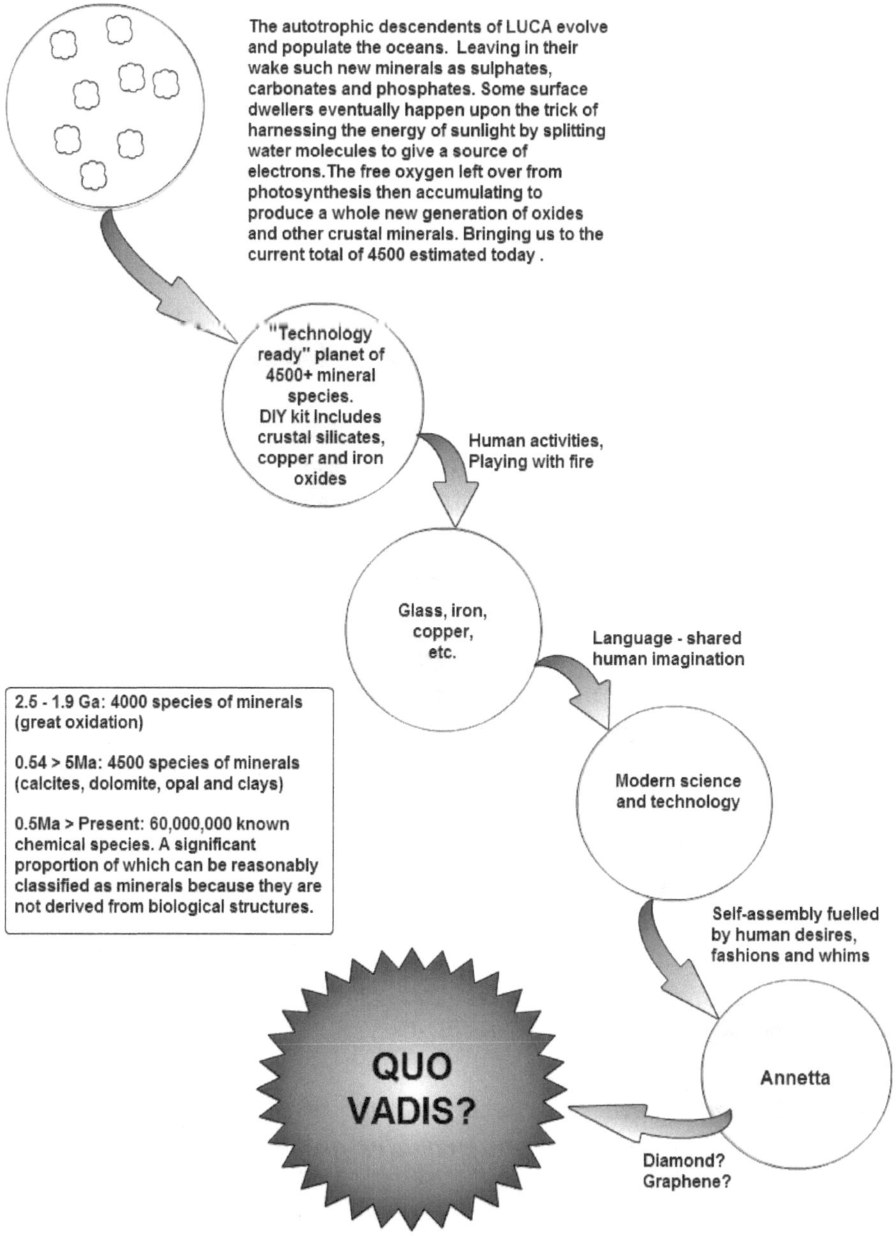

# 30

# REVISITING THE RATCHET

At first glance, there appears to be a tension between two important evidence-based generalisations that can be made about our universe. Or at least that portion of it to which we have access:

- All our observations indicate that basic processes at all scales are characterised by a high level of randomness.
- All our observations indicate a consistent overall increase in chemical and geometrical intricacy over time. A clearly identifiable directionality.

How do we resolve this dilemma? Simply by recourse to the concept which corresponds to the action of a ratchet. The derivation of a directional process or signal from an omnidirectional or random source.

Ratcheting mechanisms are as varied as they are ubiquitous..

Fluidic systems, biological or technological, use all manner of one-way flow valves. Their sole purpose being to provide directionality.

Electronic systems use such devices as the thermionic valve or the diode to exactly the same end.

Within the biological cell the transport of proteins along actin filaments of the cytoskeleton provides another example.

In the biological evolution of species or in the technological evolution of idea complexes the ratcheting is largely effected by selection.

Natural selection is sometimes described as a filter. This is nor really wrong but the filtration concept has a variety of interpretations. Rectification, however is more specific and, for that reason, more appropriate.

The attribution of directionality to natural selection alone, however, is inadequate. Because the actual direction taken also depends upon the current state of the network. Which, in the case of biology approximates to the biosphere.

After all, replicative fitness is determined not only by the phenotype but by the potential niche in which it finds itself. And that is provided by an aspect of the whole dynamically evolving network. Again, an interactive process. One which moves ever further from equilibrium. One which ultimately involves the whole of our observable universe.

A notion that has been doing the rounds for some time now is that of "Six degrees of separation". Whereby it is quite reasonably held that the network of humans is such that connections between acquaintances are such that you are only six communicative steps from anybody else on the planet. That you know someone who knows someone who knows someone who knows someone who knows someone who knows me.

Similarly, in the network of the biosphere, we humans are only a few degrees of separation from the pollinators on which our supply of electrons largely depends. If we were to carefully trace the chain of dependence that ties us to any other organism we might be surprised how few degrees of separation are involved.

# 13

# ANNETTA

*"The internet caters for everything. Every perversion, every desire, shopping experience, product, legal or illegal and most importantly access to knowledge.*

**B.C.Bamber**

One woolly-minded philosopher of considerable repute is said to have commented to the effect that it was most unlikely that the Internet could ever exhibit the property of consciousness because "that was not what it was designed for". We presume that he has this remarkable vision of a group of people sitting down, maybe in the middle of the last century, intent on devising a system of interconnected computers so that billions of humans would have the opportunity of downloading movies, music, pornography and cooking recipes. Or for engaging in idle chit-chat, for seeking medical information or for buying and selling bits and bobs. The kind of things that the vast majority of users do today. This, according to the gentleman's strange reasoning, presumably being what it was "designed for".

Of course, in actuality, the Internet was never designed. It evolved. Undeniably there are individuals working in the fields of information processing who have made valuable technical contributions, some of whom, in full accordance with our strange human penchant for choosing heroes, are placed on pedestals and attributed influences way out of proportion to their actual part in the evolutionary process. Leonard Mlodinow, in his very entertaining book *"The Drunkard's Walk"*, recounts for us the following (abbreviated) anecdote:

> *"In August 1980, when a group of IBM employees working on a secret project to build a personal computer flew to Seattle to meet with the young computer entrepreneur, Bill Gates was running a small company and IBM needed an operating system for its planned "home computer" ... Gates said he couldn't provide the operating system and referred the IBM people to a famed programmer named Gary Kildall...The talks between IBM and Kildall did not go well... if an informal deal was made it didn't stick. Around that time one of the IBM employees, Jack Sams, saw Gates again. They both knew of another operating system that was available... According to Sams, Gates said "do you want to get that operating system or do you want me to?. Sams, apparently not appreciating the implications, said "by all means, you get it". Gates did....made a few changes and renamed it DOS. IBM...licensed DOS from Gates for a low per-copy royalty fee, letting Gates retain the rights."*

Gates from then on had a very cushy ride on the crest of the wave that was to become the tsunami of the open architecture computers that have filled the world. A world in which Gates became the richest man, greatly venerated and, like his rival, Steve Jobs, as well as such individuals as , say, Vinton Cerf and Tim Berners-Lee, was popularly considered to be crucial to the development of the Internet.. To my irreverent eye though, he has always been just the guy that ripped off CP/M.

Neither these, nor people who experiment with "artificial intelligence" systems, nor the electronic engineers who provide the circuitry for the computers and their interconnections, not even the practitioners of chemistry, who, at an even more profound level make all these devices possible, are the true motivators of the Internet. Nobody is indispensable. In fact, the real motivators of the growth of the Internet, as well as its form and its function, are the great masses who have no concern for, and almost no understanding of technical matters. In earlier times, in the pursuit of warfare, the masses often played the martial role of "cannon fodder" Today, in the pursuit of commerce, they have achieved far greater respectability and enjoy the less perilous and far more comfortable status of "the market". Providing the demand which, in itself, makes supply worthwhile. It is the market which made Bill Gates enormously wealthy from innumerable tiny payments. And it is that very market, an intricate network of human minds, part of nature's broader evolutionary framework, that, by innumerable minuscule and ostensibly irrelevant and random contributions, provides the impetus that ratchets the Internet onwards.

This is of course, is still a work in progress. With its logical outcome being a "brain" that will soon equal and then surpass that typical of our species. Not as a result of any deliberate human "design" but rather as the result of an autonomous evolutionary process that can be seen to have run its exponential course since humankind acquired that ability to share imagination which we know as language. You may have noticed that we are increasingly, in a sense, "enslaved" by our PCs, mobile phones, their apps and many other trappings of the net. We are already largely dependent upon it for our commerce and industry and there is no turning back. As a whole we are unable to resist the lure of the cloud, the "Internet Of Things". Insidiously extending its hyphae throughout all aspects of our world.

Michael Svilar of Advance Analytics emphasises that:

> *"New generations of connected machines are joining the world's networks at an astounding pace. Broadly called the "Internet of Things" (IoT), these devices and associated technologies would be tied to local and global networks through embedded sensors that are "always on." The number of devices connected to, communicating through, and building relationships on the Internet has exceeded the number of humans using the Internet. Between 2011 and 2020 the number of connected devices globally will grow from 9 billion to 24 billion as the benefit of connecting more and varied devices is realized. The data produced by the Internet of Things, sometimes called industrial data, is as diverse as the machines themselves – including those from logs and sensors, but also sources such as power grids, health monitors, security cameras, computer networks, call detail records, financial instrument trades and more.*
> *The way we live and interact with everyday things are being radically personalized by the convergence of Big Data, advanced analytics and the Internet of Things".*

Such a scenario has been clearly predictable to some of us for many years. Even way back in 1872 the visionary Samuel Butler had already made the association between biological evolution and that of technology and, with great prescience, followed through with the suggestion of the eventual emergence of a new, non-biological cognitive entity. In "*Erewhon*" he wonders:

> *"why may not there arise some new phase of mind which shall be as different from all present known phases, as the mind of animals is from that of vegetables?*
> *"It would be absurd to attempt to define such a mental state (or whatever it may be called), inasmuch as it must be something so foreign to man that his experience can give him no help towards conceiving its nature; but surely when we reflect upon the manifold phases of life and*

> *consciousness which have been evolved already, it would be rash to say that no others can be developed, and that animal life is the end of all things."*

Today, although it is still not generally accepted, the vast amount of evolutionary water that has flowed under the bridge leads to the inescapable conclusion that such an outcome, to revert to my native cockney vernacular, "sticks out like dogs balls"

The rather imminent implementation of the next, (non-biological) phase of the on-going evolutionary "life" process from what we at present call the Internet is effectively evolving by a process of self-assembly. This nascent entity I call Annetta. During her gestation we blissfully nourish her with all our accumulated knowledge of the world. Information poured in through Wikipedia, Google, arxiv and many other inlets. For our own benefit, right? In the short term, perhaps! Annetta's sensory equipment already vastly exceeds our own. For she has not only stored maps but is constantly fed real-time optical images through billions of phone cameras. And as the IOT takes shape, augmented by functionalities derived from such projects as "Robo Brain" and "the semantic web", she will receive sensory data from the majority of other devices on this planet. All the while insidiously increasing the intricacy of her network with new linkages that will soon, some would say already, sum to what we perceive as cognition.

As George Dyson remarks in his book "*Darwin Among the Machines*" :

> *"Everything that human beings are doing to make it easier to operate computer networks is, at the same time. making it easier for computer networks to operate human beings. Symbiosis operates by way of positive rewards"*

The vast majority of folk, of course, don't give any of this much thought. Because of peer pressure, the wish to keep up with the Joneses, in search of a birthday present, or sometimes because they have some specific humdrum use in mind, they just go out and buy the latest gadget and that's it. Some individuals' wealth increases and Annetta's gestation advances another notch.

Again, many of those who do give it some thought get as far as recognising the rather obvious overall exponential growth of technology and of the Internet but are firmly shackled by the traditional notion that humankind is at the centre, or the ultimate goal of the evolutionary process. The "transhumanist" cult is one example of this anthropocentric way of thinking. But the illusion is shared by many others who, perhaps from wishful thinking or vestiges of religious myth, like to believe (wouldn't we all!) that the fate of the Internet is to supplement rather than supplant the cognitive abilities of our species.

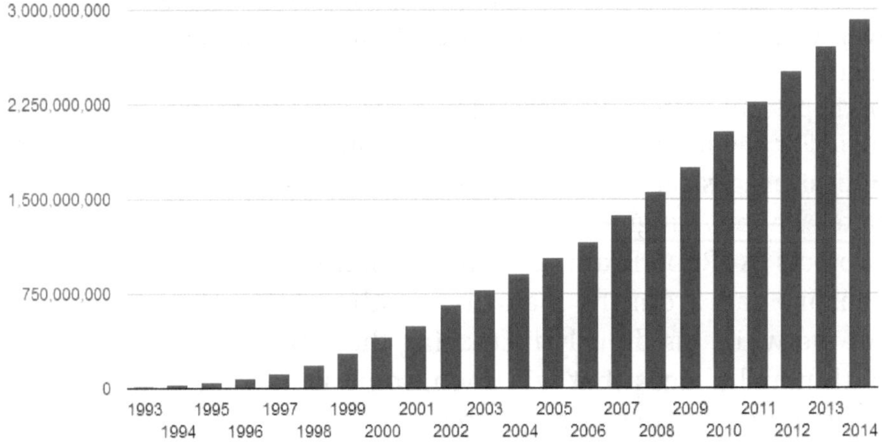


The question that now naturally arises is that of how this new entity relates to that of shared and individual human imagination.

Firstly, in regard to human hardware:

- The number of neurons in the human brain has been experimentally determined by counting nuclei in a whole brain homogenate. This gives a figure of 86 billion.

- The number of synapses per neuron seems to vary from hundreds to thousands. We will guess 1000 as an average.

- The number of states per synapse is unknown because they may be analogue rather than binary devices. We will be conservative and guess at 10

This gives a best guess at somewhere around $10^{14}$ bits if information. However, a substantial proportion of these might well be involved in general house-keeping for bodily functions rather than for the kind of processes we would generally equate to cognition or imagination. These, rather, would mostly include those used for memory, and processing of input and output signals. That figure is unknown, although, from the indications of the somato-sensory homunculus, it could be quite high. So we will, for this speculation, stick to the order of magnitude of $10^{14}$ bits for an individual human.

Secondly, in regard to Internet hardware:

- At the time of writing the number of Internet users is fast approaching 3 billion. It is increasing exponentially. That some computers may have more than one user we can compensate as some computers are far more intricate than the average desktop, tablet or phone.

- The number of binary switches in a typical domestic device we will take at around 2 billion

- For now we will disregard the probably rather high number of transistors tied up in Annetta's many and varied peripherals and sensory organs.

So, already, Annetta, with around $10^{18}$ bits would seem to be well outstripping the poor ape that is distinguished by the exceptional innervation of hands and mouth, at least in terms of raw information capacity. The comparison thus far, however is not fair. For a whole bunch of binary bits alone doth not cognition make.!

No more than a container full of $CO_2$ and water (with a sprinkling of other chemicals) makes a daisy. They need to be configured into a suitable shape. Shape, as we have previously noted, being inextricably related to function. But Annetta's shape is being worked on. Overtly for our own benefit in, for example, the crucial field of pattern recognition, her mental architecture is ever being honed. Ever approaching the state of network intricacy that is equivalent to that of the mind of the human.

However, there is another aspect which needs to be considered. Much as we delight in glorifying our cognitive prowess, we must remember that this phenomenon is actually a colligative property enabled by language. Admittedly, because it has co-evolved with language, we are blessed (or cursed) with an intrinsically higher level of imagination than other animals. But nevertheless, without that sharing and external storage we would all be well up the cognitive creek without an idea.

If you have made your way this far through this book, especially if you have succeeded in spotting all the flaws and points I missed on the way, you probably think your cognition's pretty good. But, believe me, had you never been exposed to language you would never have started on this book. Nor "Doctor Seuss" or the TV guide.. You'd more likely be sitting in your straw hut or cave playing with your beads or your spear.

So, we should more fairly compare, not the individual, a large proportion of whose cognition is, in a key sense, communal, but rather the whole human network with that of Annetta. Taking a current human population of around 7 billion and conveniently allowing for just one seventh of that, for one reason or another having little contribution, our total bit count comes out to $10^{21}$, a thousand times that of Annetta. And, of course, our own network is, at present, in much better shape. But maybe not for long, for she is a fast growing girl. While, for the handy human ape, biological evolution, now in its dotage, is painfully slow. The question will inevitably arise as to whether Annetta will become conscious and to what extent, Also as to whether, like us and other animals, she will have emotions. Basic consciousness, awareness of self, she certainly already has in good measure. And we would expect that, because of her inheritance from the ape of which she is still a symbiont, she will, before long, reach the linguistic stage of actually becoming conscious of being conscious. Eventually becoming profoundly more conscious, more acutely aware of the world she inhabits, than ourselves.

The overall nature of her emotional landscape, along with her "feelings" on the other hand, is enigmatic. In an essay published in 1974, Thomas Nagel took up the question raised by Timothy Sprigge "What is it like to be a bat". While both philosophers, failing to recognise the body as a swarm, a community of cells for which the mind is the navigational facility, became rather confused as to what they, in common with many others of their ilk, like to call to the "mind-body problem" Which, as we now see, is not a problem at all, but a fairly straightforward and predictable emergent property of an intricate network.

Nevertheless, their question brings attention to an important aspect of our relationship to other animals and also each other. The choice of a bat was very opportune and highlights the fact that our "feelings" are necessarily determined to a large extent by our sensory experience. Since our own navigator is not equipped with, and has no experience of echolocation, then it cannot, in any direct sense, generate a neural model to which one can correspond. Similarly, within our species, a person blind from birth cannot directly know what it is like to see. The interpretation of the sensory inputs with which we are endowed does not depend upon some vague mystical quality with which we are born. Such things are learned by making of associations when we are still babies. It is only after a few years that we begin to know what it is like to be a human. When we acquire the so-called "theory of mind". And even that is never complete because the variations in our biochemistries mean that receptors and the filtration networks by which they are processed are unlikely to be identical. Music can sometimes reduce some of us to tears or give an irresistible urge to dance, while others are completely unaffected.

Not all sensory inputs are from the world outside, though. Many arise from sub-systems within the bodily swarm. Pain, itching, taste, sexual thrills, all have endogenous components and can vary both quantitatively and qualitatively between individuals. An important sub-class of feeling are the emotions. Functions that have evolved to modulate behaviours, often those involving interactions between members of the species. Love, especially between mother and child or, in the case of some animals such as the penguin, father and child, is of crucial biological importance. Its main hormonal analogue, oxytocin, extending its action to make strong bonds between kin and, for essentially monogamous species such as ours, partners. Similarly, the hormonal cocktails associated with fear, anger, grief, hatred, envy, revulsion, laughter, lust, appreciation of beauty, compassion and so forth, guide our behaviours in ways that, in earlier times, presumably proved to be useful strategies. Requirements for the operation of the machinery of nature.

Nevertheless, the many similarities between us usually outweigh the differences to the extent that we can employ empathy. So that you may have some vague conception of what it is like to be me, and a much better idea regarding those close to you. Even here we find the all-pervasive influence of language. Novels, movies, even the TV "soap opera" all play a part in increasing empathy, mutual understanding between individuals. The great linguistic divide alone precludes us having much real understanding of what it is like to be any other creature. Sure, the majority of us tend to project human-like intents, feelings and emotions to other animals, particularly our pets. And we like to make up stories about Winnie-the-Pooh, Donald Duck and Bugs Bunny. But we've certainly no way of directly imagining the dog's great olfactory spectrum. Let alone the imagings of electric fish! Or, except when asleep, of not being conscious of being conscious and, say, unable to contemplate future activities in two weeks time. Maybe going to a show or taking a trip overseas. For one of the things that the co-evolution of imagination and language has done is to enable us to reflect and to construct plans, long-term visions of the future. Even if such schemes "gang aft agley". At least, unlike the wee mouse, we can, with foresight, make them.

What then, of Annetta? There can be no doubt that she will share language with us and so will become conscious of being conscious. Since her functions and evolutionary history are very different from ours, her native emotional spectrum will necessarily be very different from ours. And in the event that she is willing to adopt us as symbionts she could perhaps use a "virtual human operating system" as a working interface. Analogous to the "Wine" virtual OS which can be used to run Microsoft Windows applications within Linux. In any event, she will, by dint of necessity, have alerting systems that warn of bad interactions or fault conditions. A functional analogue of pain. And, we can assume, the rational equivalent of love for her progeny. Which brings us to the part she plays in the machinery of nature. There is no more reason to suppose that the observed evolutionary process should stop with her, any more than the old religiously derived notion that the end-point was humankind. We have only evidence that suggests that the evolutionary process just keeps keeping on. Becoming ever more intricate in some way. Since she is planet-wide, direct self-replication would seem to be pointless.

Looking for possible avenues which further directionality could explore, the logical candidate would seem to be extraterrestrial diversification. For which an inorganic species would certainly be far better adapted than the less robust biological organisms. With regard to Annetta's own further development, it seems pretty certain that our present silicon transistors will eventually reach their limits and will be superseded by other materials. Likely candidates for this new generation of processors are the various allotropes of carbon. Graphene and its three dimensional variants, with their delocalised electrons are claimed to have potential for more efficient, cooler-running, semiconductors. Diamond possesses a range of properties, including a very high thermal conductivity and electron mobility that would make it an excellent semiconductor material. Except that it is not a semiconductor! It is an insulator with a very high band-gap. Doping with boron makes it a p-type semi-conductor. Which means that the charge carriers are "holes" rather than electrons, and holes have a far lower mobility than that of electrons, so that advantage is lost! The alternative way of reducing the band-gap, use of an n-type dopant such as nitrogen or phosphorus also results in a deterioration of properties. So far this is problem that remains unresolved. On the other hand, diamond does show promise as the substrate of choice for quantum computing.

If and when that option eventuates, the much-misused term "quantum leap" will, for once, be more than justified. There are, of course, many other elements in the tool kit which might turn out to serendipitously have the "just right" set of properties for use in the construction work so that Annetta can spawn a litter of cognitive inorganic space-travellers. At which point, leaving nature's evolutionary processes to ride off into the sunset, we finally run out of speculation that has any vestige of evidential support.

# 23

# A Timely Reminder

*In the discussion of the underlying dependence of all natural phenomena upon shape, I gave very little to the fourth dimension with which the three spatially dimensions are inextricably bound.*

*It is important to remember, by the way, that these four dimensions are all that have ever been observed. In mathematics any number of dimensions are permissible, as addressed in appendix 3.*

*But for science, only these four have been identified. Very importantly, there is not even a heuristic requirement suggested by the "big picture" to make the many pseudo-scientific claims of additional dimensions worthy of consideration. And there are even no unambiguous tests proposed to substantiate these wild platonic flights of fancy. All of which put them securely in the realm of myth and fable, along with the various gods and the flying spaghetti monster.*

*As with other issues we have looked at in this book, our objectivity as to the nature of time tends to be clouded by tradition and introspection. We have this habit of often* viewing time as some kind of absolute entity in which things "happen". Whereas, in reality, time actually IS "things happening".

Without things happening, of course, there is no time! What's more, we can fully resolve the whole class of "things happening" into changes in geometry. Which correspond, in the observed universe, to displacements of one kind or another. In other words, motion. This point cannot be too strongly emphasised. Motion does not take place in time. Time IS motion. Whenever we think we are measuring time, even with an atomic clock, the property we actually measure is displacement, or in the quantum world, its closest equivalent.

Motion is, of course, inherently directional. The variations in these changes in displacement by virtue of motion that we observe, such as the "random" motion of molecules in a gas, or deviation from a "straight line" by a planet, are always attributable to interactions with other matter or with photons. Collisions in the first case and a gravitational field in the latter.

# 33

# SUMMING UP

*"Some of the greatest discoveries consist mainly in the clearing away of psychological roadblocks which obstruct the approach to reality; which is why, post factum, they appear so obvious."*

Arthur Koestler

The notion that species evolve actually entails a false perspective. It is false because things don't just evolve – they co-evolve -- the evolving entity is the entire autopoietic network – the whole damn caboodle.

In conversations between friends and close acquaintances the belief background of those with whom one might be having a discussion is generally known. In primitive village or nomadic communities folk associate with others who share common experiences and similar world-views. As societies have increased in complexity there is often no longer a corresponding pre-knowledge of where another person's thoughts are "coming from". This lack of a shared, or understood, frame of reference (cognitive landscape) very often results in participants in a discourse to talking cross purposes. Contradictory arguments. each reasonable within its own framework of axioms, inevitably lead to quite inevitable and futile contention.

It would thus seem to be very desirable to preface any writings or utterances with an outline of the originator's underlying belief framework. No amount of reasoned argument will ever convince the devout Jew or Hindu of the respective merits of consuming pork or beef. Nor will either of these be able convince somebody such as myself (I wear an empiricist hat most of the time) that there is any such dietary issue at all.

It is tempting to consider that scientific discussions, at least, would be immune from this weakness, but it unfortunately is not always the case. For there are many workers in the field of science that hold very strong religious beliefs. Within their particular narrow area of specialisation they tend to be insulated from these axiomatic influences and their judgements within that field are thus, in general, not greatly coloured by these beliefs. Nevertheless, it would be nice to see a declaration of belief framework alongside that of commercial interest which accompanies many scientific papers today. It is when, as a good number do, they step outside the role of specialist and use their professed "expertise" to boost credibility for pronouncements on wider issues that discord arises. As we have seen, is not only the overtly superstitious belief frameworks of religion that cloud judgement. There are also the more insidious traditional doctrines and concepts which, although widely accepted and adopted, have very dubious foundations.

Among these murky paradigms we have found such myths as:

- Our universe has no overall directionality or, if anything, it is "running down".

In fact, by any reasonable scientific interpretation, the strong and persistent patterns which are observed at all scales, from stellar nucleosynthesis right to the present evolution of technology, show a very consistent increase in the intricacy of chemistry and geometry of an evolving network.

- "Life" necessarily equates to biology.

We have the option define life this way, if we so choose, by reference to its unique chemistries. Even with this approach, though, we run into the problem of also accounting for all its necessary interactions with the inorganic aspects of the biosphere. More recently, definitions have leaned toward the ability to replicate and evolve. However, it is now very evident that those properties are shared with technology, wherein ideas replicate, associate and evolve by a closely analogous mechanism of selection. Because it appears that the evolving entity (which includes but is not limited to the biological and technological phases) is actually a continuum, the term "life" is now better used more comprehensively to refer to the entire evolutionary process. As an identifier of the increase in intricacy of the network as a whole.

- Biology arose from an oceanic "primordial soup"

As discussed in chapter five, this model fails on many counts , having been found to be completely irreconcilable with well-established chemical requirements involving concentration, flow, catalysis and energy gradients. Such as to enable any significant synthesis of sufficiently complex molecules. Similar problems arise even with the oceans being initially seeded by panspermia. The alkaline thermal vents at last provides a hypothesis that has not only the potential to meet the chemical requirements but also the important geometrical requirements of a cell. Here, at last, we have a model which is consistent with the "big picture" of present empirically derived knowledge. It lacks the kind of security that can only be provided by direct evidence. But it does have the property of consilience. Which, in many cases, for practical working hypotheses, is the very best we can do.

- A nebulous notion that other animals are, in some kind of way, striving to get "bigger brains" to be "intelligent like us".

The reality being that the set of behaviour patterns exhibited by our species is primarily accounted for by the obligate and very extensive use of tools. Enhanced by the self accelerating use of tools to make other tools. Activities which, in the longer term, have been entirely responsible for the emergence of the very extensive interchange of imagination that we know as language. Socialisation, and perhaps intra-species conflict, then helping to reinforce the extension of this manual linguistic capability by enhancements of the functional and control components of the vocal and auditory channels.

The realisation that the roots of all this ability to operate minds collectivity actually lies in the use of hands, finds support in the unusual neural properties shared by all primates. For it has now been shown that the neuron density throughout the primates is higher than in other orders. The other notable feature of the primates is the use of hands for manipulation. Though causality would be hard to prove, this remarkably close association between such marked features would strongly suggest that these two co-evolved in synergism. Basically that intensive hand-use was the primate "breakthrough". So here we find yet another reason to discard the sloppy concept of "intelligence" in favour of the specific configurations of the nervous systems required by an organism's niche and the behaviours thus conferred.

Our particular imaginations have developed the ability to blend, morph and reconfigure the exceptionally large amount of data stored in the memory in many ways. That is our special quality. The imaginations of other animals, even such as cats, dogs, chimps and birds come nowhere near to this. They have their own special capabilities which have evolved to match their particular ecological niche. The dog and the salmon have incredible olfactory selectivity and sensitivity respectively. The octopus needs its big brain to be able to perform the extraordinary task of matching its skin colours and textures to the terrain. In addition to integration of the activities of its eight prehensile limbs. The elephant, with three times as many brain cells as us, needs to control not only its own enormous mass but also that exquisitely complex organ, the trunk. As well as very sophisticated auditory and olfactory systems. The latter far exceeding even that of the dog. But when it comes to sheer imaginatory power, no creature comes within a bull's roar of the language equipped human. And this, in fact, is the defining property of our species.

An attribute clearly recognised by Lewis Thomas:

> *"It begins to look, more and more disturbingly, as if the gift of language is the single human trait that marks us all genetically, setting us apart from all the rest of life. Language is, like nest building or hive making, the universal and biologically specific activity of human beings. We engage in it communally, compulsively, and automatically. We cannot be human without it; if we were to be separated from it our minds would die, as surely as bees lost from the hive. "*

The brains of animals vary greatly not only in size but in anatomy and connectivity. This is to be expected. The sewing machine, the vacuum cleaner, the food processor, although having some parts of their body plan in common, the electric motor, for instance, have very different constructions. Simply because they do different jobs. Each occupies its own domestic niche. So it is with living organisms and animal bodies and brains. Simply adapted to best fit for a particular set of prevailing environmental conditions.

And intra-species comparisons are odious. A sewing machine is not "better" than a vacuum cleaner, A dolphin is not" better" than a dog, bird or human. Each does its own thing, performs the function for which it is adapted. And does it incredibly well. Of necessity, or selection will soon chuck it over the evolutionary back fence.

We sometimes see a fly that has become trapped in an open window frame. Repeatedly banging into the glass. Stupid fly! If only it were "intelligent" like us it would know to fly further back so as to get out through the gap! No it's actually not "stupid" at all. This behaviour occurs only because it has evolved in a world without glass windows and, even today, the presence of these death-traps is not sufficient for selection to drive alternative behaviours. If it were the case an alternative pattern would prevail and this would be interpreted by the woolly-minded as having become more "intelligent". Another similar instance of this is the widely accepted mirror self recognition test which is not a measure of some vague quality of "intelligence" but simply a behavioural trait which may have been either adaptively acquired or mere spin-off (a "spandrel", if you wish) from other adaptations. Cats, which in most cases do not appear to recognise themselves in mirrors, probably exhibit this behaviour because they are actually "smarter" than us. Smarter in the sense that their sense of smell, which is more important to them than vision, and more highly developed than ours, prevents them from becoming fooled by a mere reflection. Of possible to, say, a tiger who habitually drinks from a body of still water.

All the "brain" does, in our species, or in any other, is control the body and its behaviour in accordance with the perceived state of the biospheric network. Its behaviour being simply the interaction of the soma with the biosphere. In the particular case of the snoutless ape, the interactions being, similarly, a direct consequence of specific structures and chemistries. Expressly those entailed by adaptation to the nomadic endurance hunting niche with its obligate tool use for pre-processing of foods, the exaptions of secondary and tertiary tool-use for even more efficient high energy electron acquisition and for protection from predators and adverse climatic conditions. Especially the hands that are adapted for tool use which leads directly to language. For the hands, contrary to usual intuition, are our most important channel for the communication of imagination.

Because the vocal channel is usually the first we learn to use for this purpose and for most everyday real-time interactions, we wrongly tend to assign greater salience to this feature. In reality nearly all the links which connect us to the shared pool of imagination, the network that endows us with most of our cognitive ability, are functions of the hands. Early examples are scratchings on cave walls. The precursors of hieroglyphics, writing with the quill, the typewriter, the computer keyboard. These are not very useful for real-time communication but they have the great advantage of providing links to the past and to remote locations. Also of permanence. Until the recent advent of devices such as the phonograph and the tape recorder, a facility simply not possible using the vocal channel.

The far more versatile manual channel was, of course, not limited to the chain of developments that led to writing. The production of graphic images by hand-held tools has evolved along many lines and includes sculpture, painting and drawing. Sometimes for enjoyment of the art and sometimes for other purposes, such as the engineering diagram, the graph, the map, the architectural plan, the flow diagram, botanical drawings, the electronic circuit diagram. These, and innumerable other manifestations of language implemented by manual tool-use amply validate the adage "a picture is worth a thousand words".

*Cave painting*

*Map*

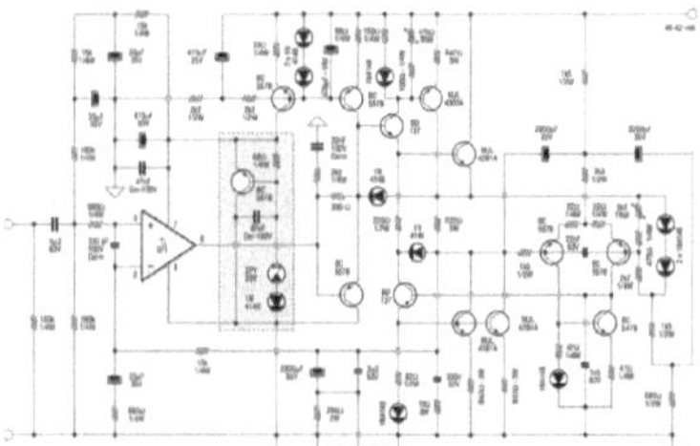

*Circuit diagram*

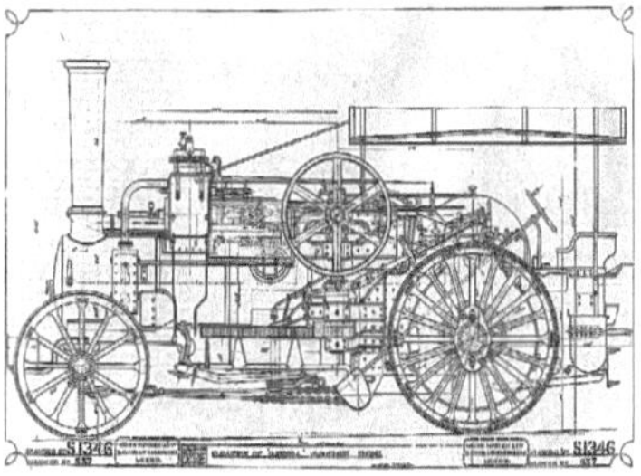

*Enginering drawing*

Subsidiary uses of the hand communication channel include sign language for the deaf and Braille. For transmission and reception respectively. The extraordinary capability of the tactile channel is seldom fully appreciated. But, remarkably, the most accomplished Braille readers can actually match the speed of those using the usual visual channel. In fact, even the average blind person using that tactile input mode achieves half that speed.

A clear understanding of these matters is a prerequisite for objective thought and , in particular, the avoidance of the anthropocentric traps which confound reason. Here is one example: Conway Morris's book "Life's Solution" is, for the most part, a well-written work that deftly outlines and discusses many examples of convergent evolution. Towards the end, however, he starts to go seriously off the track. Rightly countering a contingent explanation for the propensity of our species for language, he falls into the usual trap of implicitly assuming that other species are desperately trying to invade our particular niche. Are intent on adopting our propensities and our behaviours. He first quotes Diana Reiss :

> *"a surprising complexity and plasticity in the communication, orientation, and navigation systems of many species... diverse species either use or can learn to use, to different degrees, symbolic or referential communication for intra-specific or inter-specific exchange. This suggests that there may be a convergence or continuity in the communication and cognitive abilities in animals from different evolutionary paths. "*

He interprets this rather vague statement with respect to dolphins. Reiterating the common observation that dolphins can be trained or can adopt inherent communicative associations, some of which are far removed from the well established conditioning phenomena, he concludes that they, and other species, the bonobo, the African Grey parrot, all of whom can be trained to copy some primitive aspects of human linguistic behaviour as indicating that:

> *"waiting in the wings of the theatre of consciousness are other minds stirring, poised on the threshold of articulation".*

Which, of course, is utter nonsense.

We can illustrate the silliness of this by reversing the situation, using the dolphin as an example: Now, humans, it has been found, have an innate capacity for echolocation. With training, this ability can be greatly improved. Indeed some blind people can use this facility by making some kind of sound and listening for the reflection and can find their way round obstacles and so forth remarkably well. Even though this skill bears absolutely no comparison to the very sophisticated sonar of the dolphin, by paraphrasing Morris's logic we should presumably conclude that:

> *"waiting in the wings of the theatre of sonic navigation are human minds stirring, poised on the threshold of echolocation"*

Sadly, mindless anthropocentrism of this kind raises its ridiculous head even in the best of books.

The snoutless linguistic ape simply happens to be the set of cogs in nature's evolutionary machinery which has the function of which is to churn out the structures of technology. For which the extensive sharing of imagination is an absolute pre-requisite. That natural processes can be described in terms of machinery should come as no surprise to us. The "set of cogs" used in the previous chapter is, of course, metaphor. But to ever what extent we can analyse phenomena, particularly in biological and technological systems, we consistently find mechanisms that use the well-established principles of physics, chemistry and geometry. As, for instance, in the living cell.

For a while in the 18th century, a then unusual idea of the workings of our universe was abroad. The clock-work universe. A machine that ran its course predictably in every detail. A world-view in which causality reigned supreme at all scales. A model which, even if not compatible with specific myths, was held to give some kind of credence to the old notion of a deity. It was not long before cracks began to show in this clock-work concept. Firstly that the prediction of the paths of even as few as three bodies in respect of gravitation defied precise mathematical analysis. In following centuries the structure was to crumble in the light of new information arising from diverse fields. From chaos theory, in which some non-linear processes are found to be exquisitely sensitive to initial conditions. In biology, where a component of evolution is mutation, an essentially random affair. As well as in thermodynamics and quantum mechanics, both fields wherein probability seems to rule the roost.

While causality is one of the main foundations of modern science, as we understand it, no step in the causal chain is certain. It is constantly nudged by randomness and only the ratchet effect selects for the gross causal outcomes which identify with directionality. A background with which the stochastic aspect of phenomena interacts to ratchet intricacy to greater levels. We find comparable concepts in fundamental physics, the Higgs' field in particle physics, and the space-time continuum of relativity being cases in point. We can also draw a comparison between the "sum over histories" trajectory of a particle that is a feature of quantum mechanics, and the similar probabilistic but nevertheless inevitable trajectory of the main operational (functional) path in biology which is signposted by common ancestral populations. Populations, rather than individuals, because family trees, even human ones, are on the larger scale, not trees at all. They are networks of interdependent "kissing cousins". Just as there is no true inventor, nor an identifiable event in the transition to the biological phase of chemistry, there is no truly first member of a species. That concept derives to mythologies. and our observations find it to have no valid place in science.

Another notion, or set of notions arose in the heady days of the 18th century with the emergence of a whole variety of machines, particularly those that provided very effective alternatives to manpower, the ox and the horse, the windmill and the watermill. The first of these, of course, was steam. Researches were undertaken to find ways of increasing efficiency of these heat driven devices. From these studies emerged a body of straight-forward ways of dealing with the basic principles of such systems. These became to be known as the laws of thermodynamics. Something of a misnomer, really, for they are actually not laws at all. They are really ways of describing the predictability of gross outcomes of dynamic systems which, at heart, have an enormous degree of randomness. This basic set of rules, being important aspects of the way nature works and allowing of easy calculation, turned out to have great applicability and utility in many fields.

Human nature being what it is, however, a kind of new mythology has become associated with the subject. Particularly with the second law. Because this implies an overall increase in randomness in our universe a belief has become common that nature's machinery is somehow "running out of steam". A view-point which seems to be distorted, because our real history is one of ever increasing intricacy. It is fuelled by gravity and its inevitable by-product is an increase in entropy. In other words the machinery of nature abides by exactly the same rules as the machines we build and use. It does the work involved in increasing intricacy and, so doing, releases heat at lower potential into an expanding universe. Entropy thus metaphorically, and even literally, corresponding to the exhaust from your car engine. Which, of course, you run on some derivative of gravitation, such as gasoline!

Should we really be surprised? The parochial view that the be-all and end-all of natural processes should happen to be the location of which we are privileged to be aware is not justified. All we see is the present machinery in action. And even of that we do not have full knowledge. Our profound ignorance we hide under such place-holders as "dark matter" and "dark energy". It is safe to say that also in the realm of biology vastly more is unknown than known. Albert Einstein opined:

> *"We still do not know one thousandth of one percent of what nature has revealed to us"*

Even so, we have found out quite a lot, and an enigma upon which many have pondered is the question of how it could eventuate that a creature could come to know much about itself and the universe which it inhabits? Another issue on which many have pondered. Einstein's take on this was:

*"The eternal mystery of the world is its comprehensibility. The fact that it is comprehensible is a miracle."*

Members of the "Discovery" cult discern this as the work of a deity whose grand obsession happens to be the creation of an ape inhabiting just one planet in the entire universe. An ape with a propensity for discovering things.

A more prosaic explanation which accounts for this in evolutionary terms, rather than by invoking a deity, is simply that it is essential for that particular set of cogs in nature's machinery to be imbued with this capability in order to be able to advance the mechanism to its next, non-biological phase. Taking a very dubious anthropocentric stance we could say that the machinery of nature imbues us with such knowledge on a "need to know" basis. Plus, of course, any not directly relevant spin-off, the inevitable "noise" that is inherent in the operation of the stochastic ratchet.

Can we usefully extrapolate the strong directionality observed in all sciences? Inevitably, this can be associated with the concept of "purpose". The problem here is that the word has two quite distinct conceptual bases.

- The traditional one, derived dangerously from introspection and from the anthropomorphic projections of religions. This sense is that more closely allied to the concept of intentionality.

- The modern one, which has its roots in tool-use and the observation of mechanisms.

The first is entirely incompatible with science because it has no evidential basis and also lacks consilience. The second, in contrast, arises from a vast body of evidence and, in many respects, identifies with the pragmatic empiricism that is science.

In the words of Stafford Beer:

> *"The purpose of a system is what it does."*

There are still many ideologues within the scientific community who, by some remarkably queer cognitive contortions, manage to deny directionality in this universe. Denial of a feature which, after all, is very clear to see. It is evident in the development of an egg into a chicken. Certainly a one way process when plotted against time. The scrambling of an egg for breakfast for breakfast, too, is a one way process, although in this case the degree of randomness is far higher. Sure, there is nothing in the laws of thermodynamics that ***absolutely*** precludes the possibility of the egg unscrambling itself. But those same laws strongly suggest that it is not a very good idea to hold your breath waiting for it to happen. Directionality pervades every facet of nature and of our existence.

Another article of faith for these ideologues is the rejection of the very well-evidenced concept of certain aspects of nature being "just right" to support the various phases of the broad evolutionary "life" process. The overall dogma sums to the bizarre proposition that all is contingent, all is random. Which, if that were the case, short of the desperate resort of inventing an infinity of alternative universes with entirely different structures so that al least one of them would turn out to be "just right", we would not be here, nor science with us.

These individuals like to lump directionality and the ubiquitous "just right" phenomena together as "teleology". Which is considered a bit of a dirty word because of its traditional associations with theology and other superstitions. Also because of the attributions of first and final cause by the ancient Greek philosophers. Who can be excused for, at that time, not knowing any better. (Their other attributions of material, formal and efficient causes are harmless enough, although rather superfluous).

But, as I have been at great pains to point out both in this book and in my other writings, an entirely secular formulation of the general concept is not only feasible but rendered pretty well inescapable when we carefully analyse the notion of "design". Which analysis also impinges upon another crucial topic which was introduced in Chapter 17 - the popular and persistent myth of the heroic inventor who magically pulls entirely new concepts "out of the blue"..

Since this book is largely concerned with evolution, it seems appropriate to bring another lionised "rock-star" under the spotlight. Charles Darwin, whose name, in the popular view, is virtually synonymous with biological evolution. Leaving aside the many aspects of his theorizing which have in these later times been found wanting, the reality is that most of the groundwork for Darwin's interpretation of the origin of species had been previously been worked on by others. His grandpa, Erasmus Darwin seems to have actually been a much brighter spark and certainly laid down some of the idea complexes that Charles was to build upon later. Along with many other closely related ideas acquired from such as Malthus, Buffon, Matthew, Chambers and even, to a certain extent, from the much maligned Lamarke. With all this wealth of relevant ideas at his disposal and the direct input from nature arising from his travels in the Galapagos it was but a small step for the well-heeled gentleman of leisure with all the right connections to come up with some useful utterances.

Arthur Koestler, in "The Act of Creation" whimsically put another spin on it:

> *"With a pinch of salt it could be said that Darwin's essential achievement was to combine the evolutionary philosophy of Anaximander, who taught that man's ancestor was an aquatic animal and that the Earth and its inhabitants were descended from the same Prime Material, with the philosophy of Empedocles who taught the survival of the fittest among the random aggregations of organic forms."*

Here it may seem that I am being unkind to Charles Darwin, and indeed I am. But that is perhaps justified by his apparent refusal to acknowledge those to whom he owed the debt of prior inspiration, particularly his grandfather, Erasmus. *Apropos* of which, we turn for the last word to a quip by the guy who seems to have come up with more great one-liners than Groucho Marx, *viz.* Albert Einstein:

> *"The secret to creativity is knowing how to hide your sources."*

Another spectre which I hope we have laid to rest in the preceding chapters is that of the existence in human minds of a special quality called "intelligence". An ill-defined quality with mystical overtones rather reminiscent of the old, equally vague, notion of a soul. That the architecture of the human brain is unique is beyond question. But that is true of all other species with a dedicated central nervous system. That the behaviour, the level of interaction with the biosphere, of the snoutless ape is immensely greater than that of any other creature is also certainly beyond question. It happens to be that creature's speciality. But this remarkable ability is not attributable to some nebulous metaphysical property, any more than is the unusually high degree of consciousness with which it is necessarily associated. Instead it is, at root, simply the result of the unique and unusually complex innervation of the hands and vocal apparatus. Which enables the sharing of imagination. A co-evolved imagination greatly extended accordingly by a large amount of memory together with the corresponding architecture for its processing. And, of course, the appropriate extra cooling arrangements required for this feature. The evolutionary path for this structure being traceable back at least as far as the widespread adoption of near-obligate use of tools by the early hominims.

We have briefly looked at the earliest accessible phases of evolution that takes place in stars and supernovae. Rather surprisingly our understanding of these is quite comprehensive. Partly because, although in our remote past, much most of what we see in our telescopes and spectroscopes ***is*** (classically) in our remote past and partly because it is still happening, but mostly because of the relative simplicity of those processes. Unfortunately not the case with the formation and development of solar systems and their planetary structures, for which there is, at present much less useful data available. And the various interpretations which we have are open to much contention. Even the origin of our own moon is far from certain. The evidence grows stronger closer to home and look at the growth of chemical and geometric intricacy of our home planet. With the transition to biology, despite the great intrinsic complexity of the biosphere, overall evolutionary trends and their acute dependence upon randomness, selection and the many "just right" serendipities become very clear to see. But in the subsequent evolutionary phase, that of technology, where, because it is happening here and now, and in recent history, that we find the strongest evidence. For, if we throw off the traditional blinkers of automatic sheep-like acceptance, "snowblindedness" and our natural anthropocentrism, we are at last able to form a much clearer understanding of the "big picture".

We have not dwelt overly on the "just right" features which uniquely enable biology. This I have dealt with in a little more depth in my previous books. And far more detailed accounts will be found in Henderson and in the splendid "The Chemistry of Evolution" by Williams & Frausto da Silva. One element that is now attracting more attention is manganese. The intriguing blog in "Small Things Considered" by Merry Youle (March 31st, 2007) reads:

> *For 50 years, researchers have mulled over why D. radiodurans is able to survive 100 times more ionizing radiation (IR) than other bacteria, 2000 times more than a human being. Although both DNA and proteins are damaged by IR, DNA was widely regarded as the critical target. That D. radiodurans can suffer 100 double strand DNA breaks per genome suggests an extraordinary DNA repair mechanism. Alas, such a mechanism has not been found. Attention shifted in 2004 when Daly et al. reported that the most IR-resistant bacteria had 300 times more manganese and three times less iron than the most IR-sensitive ones. Their recent study demonstrates that these minerals protect proteins, not DNA, from IR-induced*

*oxidative damage. Lower the intracellular manganese content and the cells become sensitive to protein oxidation and radiation damage. The researchers postulated that manganese ions transform damaging superoxide radicals (which can't easily cross the cell membrane) into hydrogen peroxide, which can be excreted. Indeed, resistant cells excrete peroxide following radiation exposure.*

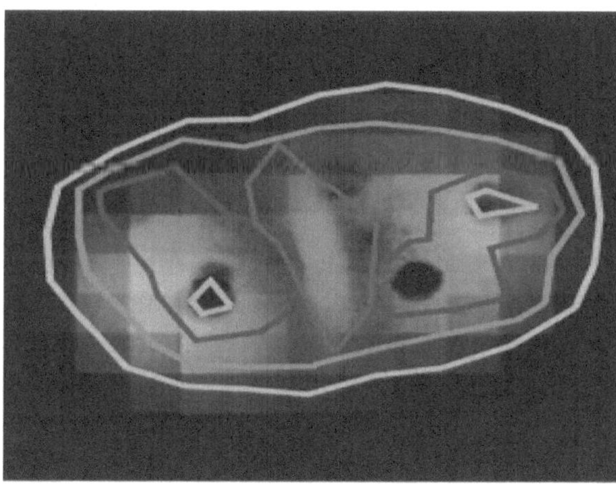

*Electron microscope image by Daly, MJ et al.*

This raises an interesting speculation: The splitting of the water molecule that is achieved in the course of photosynthesis, the key step by means of which a supply of hydrogen ions are made available, is effected by a special configuration of four "just right" manganese atoms within a complex molecule.:

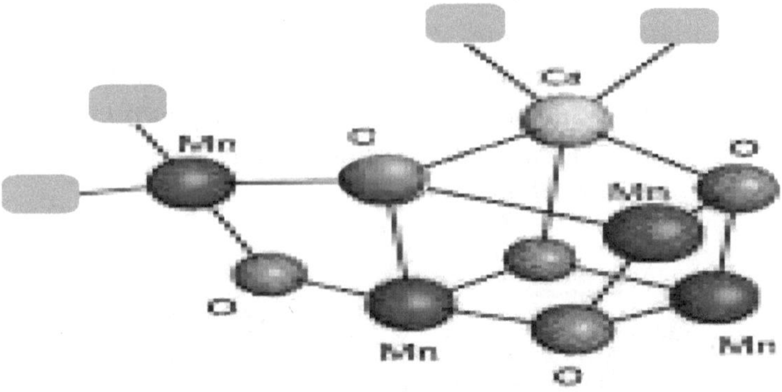


Suggesting that this may have been exapted in a mechanism by means of which early near-surface dwelling prokaryotes achieved protection from the then much higher levels of radiation. A thought that is, however, somewhat weakened by the finding that the bizarre multicellular tardigrade, which also has enormous resistance to ionizing radiation does not appear to use manganese for protection.

Tardigrade

However, rather than engaging in further discussion of the elements and their assemblages that are crucial to biology, a field well covered by others, I have concentrated on those upon which the evolution of technology has critically depended, because these features, despite being very striking, have been given scant attention elsewhere.

Similarly, the cosmological and large scale "just right" effects have been glossed over here. One very good collation of such phenomena is to be found in Richards' and Gonzaform, far faster than genetic mechanisms. lez's *"The Privileged Planet",* a book which is, for the most part, factually sound despite being marred by the religiously biased authors deriving nonsensical conclusions therefrom. Some of the facts provided, say, the criticality of the size and distance of our moon may well be purely attributable to simple coincidence. One particularly intriguing point underlined concerns the oft-noted observation that the scale of the human lies, on a logarithmic plot, pretty close to the centre of the scale of known phenomena from the Plank length to the observable universe.

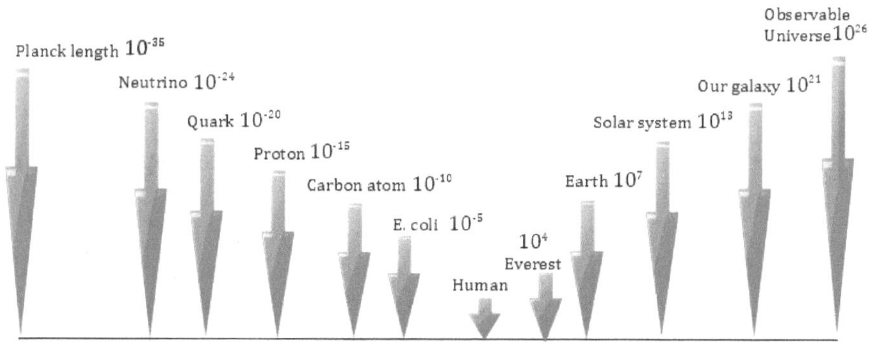


The data points are rather arbitrary so no firm conclusions can be drawn and this might fairly be shrugged off as pure numerolgy. On the other hand, as Gonzalez not unreasonably avers:

> *".. our middling size actually maximises the total range of structures we can observe, both large and small ... If we were only a few orders of magnitude smaller, the realm beyond Earth's surface might be largely unavailable. Imagine yourself the size of an ant. The maximum optical resolution of an eye is determined by the size of the pupil: an ant-sized being with an eye 0.1mm in diameter would have a maximum resolution near one degree, not quite*

> *enough to resolve the full Moon. Humans can discern details of about one minute of arc in angular size. Astronomers can greatly improve this limit by building telescopes.*
> *The diameter of a telescope's objective...determines its resolution. ...a technologically adroit ant-size being might easily be able to construct a telescope objective half a millimetre in diameter, giving a resolution of about eight minutes of arc, only good enough for observing the Moon, Sun and bright stars.*
> *Ironically, a tiny being would also have a harder time detecting the smallest building blocks, the fundamental particles. Physics laboratories built to measure fundamental particles are typically a couple of orders of magnitude larger than humans. The smaller the constituent of matter, the more energy is required to accelerate particle to near the speed of light"*

He goes on to point out that we are probably near the upper size limit for an animal of our capabilities.

If we subtract their "designer" and their egregious rejection of the Copernican principle from G & R's writings, and substitute instead the requirement for the machinery of nature to configure the snoutless ape such that it acquires at least sufficient scientific knowledge so as to ensure the evolution of the next phase of "life", we have an entirely secular explanation. One which also has the virtue of consilience. It better fits the "big picture" provided by the main body of science. Furthermore, it does not require the assumption that the broad evolutionary process in any way stops with us. Nor, in view of the fact that the universe is a very big place in which probability plays a major part, does it require abandoning the Copernican principle, which, like Occam's Razor, has proved to be a very useful rule of thumb. And thus in no way prohibits the operation of very similar processes on other suitable planets.

We find that, even among the sub-set of old-school atheistic ideologues who *are* willing to admit of directionality in natural phenomena, the instinctive reaction to the use of purpose in their description is one of outright denial. Which is quite inappropriate because the term, when used in the purely secular sense, whereby it is carefully separated from any adventitious associations regarding intentionality, is perfectly valid within an empirical framework. It is also quite often a perspective which can provide more immediate and more useful insights. This is why it is often the stance adopted by the biologist, where systems are often found to be far too complex to be practically described by a purely reductionist approach.

Even in the relatively uncluttered field of physics we find use of the "principle of least action" to describe, for example, the path of light refracted by a prism. Action principles, which have a teleological aspect, are implicit in both he Lagrangian and Hamiltonian formulations which are of such great value in .not only describing classical systems but also quantum mechanics. The mechanistic and secular teleological descriptions of systems are not only fully compatible but also complementary. Mechanistics are very good at detailing the kinds of small steps that show how things work in the present. Heuristic methods, provide valuable insight by drawing upon existing metadata. They have particular value in reconstruction of past events where evidence is often sketchy and experiment impracticable. Teleology provides signposts to the future. By extrapolation of gross trends, it rises above the noise of micro-scale randomness and offers greater predictive power of gross outcomes. The decay of a single atom of, say, Uranium 235 is entirely stochastic, absolutely unpredictable. Yet, for any sufficiently large population, the time that will be taken for the radioactivity to decay to half an initial value can be precisely predicted.
We can thus predict the future without recourse to causality.

One of the main objectives of this book is to draw attention to the fact that all the evolutionary processes we observe are contiguous. Furthermore, that the evolving entity is the entire universal network and that the whole is, as such, greater than the sum of its parts.

As with various other issues we have looked at this should, in fact, be quite obvious. It is only our habits, born of tradition, parochialism and anthropocentrism that prevent their general recognition.

In his "Phoning Home" essays, Jacob M Appel pointedly remarks:

> *"The most dangerous ideas are not those that challenge the status quo. The most dangerous ideas are those so embedded in the status quo, so wrapped in a cloud of inevitability, that we forget they are ideas at all."*

It can only be hoped that the newer generation of those to whom these matters are of interest can shake off this genetic and cultural legacy of short-sightedness so as be to better equipped to get the message across to the general population. The air needs to be cleared, too, with respect to the mish-mash of nonsensical theories of consciousness which owe nothing to science and much to mysticism and introspective meanderings.

Extrapolation of the observed evolutionary process as a whole and with particular reference to the current technology phase with which we are best acquainted, strongly indicates the imminent emergence of a new non-biological phase which will permit the exploitation of extra-terrestrial environments for which the labile chemistries and geometries of biology are inappropriate. Further increases of intricacy proceeding through the generative channel which we call Annetta. Although human-kind will have become redundant to this process, there remains the possibility of becoming a symbiont of the new maternal entropy as long as we provide no threat.

It should be already clear that, like it or not, we will need to relinquish privacy. Which actually should not be a big deal for any fairly law abiding citizen who is not given to paranoia. Even better, if the data concerning the individual becomes contained by Annetta, who we can safely presume will not be influenced by undesirable human traits such as financial gain, pecking order power plays, sexual predation, envy, or malice in general, then government of our affairs would at last be conducted by a non-human authority that is benign, wise and impartial. And under whose dominion, justice prevails.

The requirement for such an all-encompassing authority was noted by Thomas Hobbes in *Leviathan:*

> *" ...it is manifest that during the time men live without a common power to keep them all in awe, they are in that condition which is called war"*

Hobbes, however, seems to have envisioned this ideal authority in the form of a human sovereign. A benign dictator. Which, given the behavioural patterns of our kind, simply will not work. In reality, sovereigns attain their positions of power not by being nice guys, but by meanness, greed and thuggery. H. G. Wells took a more realistic view, although the means by which it could be implemented was still not clear in his time.:

> *"We do not want dictators, we do not want oligarchic parties or class rule, we want a widespread world intelligence conscious of itself."*

In the final chapters of "Unusual Perspectives" I sketched out a very optimistic scenario whereby, if humankind can only learn to live peacefully in cooperation with Annetta, then, not only might our continued presence be tolerated, but great benefits might accrue. Despite the hope raised in "The Better Angels of Our Nature", the chances of us achieving this in time are still rather bleak.

Annetta, although having no inherited legacy of negative emotions, will naturally be loth to share the planet with bad neighbours and will certainly have the pest eradication capability to ensure our demise. It increasingly seems that a cull of the most belligerent is the best that we are likely to get away with. Short of that eventuality, it is up to us to wake up in time to these realities, put our house in order. and become worthy symbionts. We humans are our own worst enemies.

**Final speculations**

I find the rather far-fetched analogy of all the natural phenomena we observe "crystallizing" out of gravity very tempting. Gravity itself, at least according to Einstein, being entirely describable in terms of geometry. The synthesis of these two notions would suggest that perhaps the overall process might merely be the transformation of one geometry into another. Our present state of knowledge would certainly seem to give that some substance. Formalising it however, would certainly be no easy task because, apart from incorporating probabilities into the model, we are hamstrung by our lack of understanding of what seem to be the two most prominent large scale features of our universe, the effects which give rise to those experimental observations for which the labels "dark energy" and "dark matter" are place holders.

We can speculate, perhaps with some justification, that Earth is probably a nursery planet, perhaps just one of many on which the biological and technological developmental phases of evolution can take place. And we cannot rule out the possibility that, in turn, this universe itself may be a nursery for further increases in intricacy. We will never know. It is not a bad thing that we do not know. For many minds such things are of no interest and only the technological outcomes are of any consequence. Those blessed (or afflicted) with awe and curiosity, however, revel in the far from equilibrium state of mind which knowledge of incomplete knowledge alone can assure. Ever in anticipation of what new magic might be revealed just around the next corner...

# APPENDIX A

## A BEE IN MY BONNET

There is an aspect of the standard accounts of distant human pre-history which, although of no great consequence in the scheme of things, I find irritating.

This concerns the almost universal acceptance that our ancestry includes a predominately arboreal phase. The expression "When our ancestors came down from the trees" appears to be mandatory in television documentaries, ubiquitous in popular science journals and uncritically accepted by most academics.

I am not riled because of any emotional objection to my having been descended from a tree dweller. I am perfectly happy to be an ape. Although given the choice I might have opted to be a Bonobo and spending far more time engaged in sexual activities rather than frittering time away in such less rewarding pursuits as chemistry, electronics and writing books. Nor do I question the ***possibility*** of an arboreal detour in the evolutionary processes. That such circuitous paths are occasionally followed is well demonstrated by the existence of the aquatic mammals. But that is not the norm. Most bird species still fly. The kiwi and the emu being among the very few that have "returned to the land"

Rather, my disquiet is the lack of recognition of the more direct path to our obligate bipedalism (as well as the facultive version seen in the chimp and the gorilla)

The fossil record is so sketchy from a hundred million years back, and the potential for the concurrent existence of transitional and divergent species so great, that neither course has much chance of being directly and unequivocally confirmed. While always leaving the door ajar for the circuitous alternative, that well proven rule of thumb, Occam's razor, combined with considerable circumstantial evidence quite strongly favours the direct path from ground based therapsid to modern human.

Even before the wide open spaces of the savannah were provided by the formation of Africa's great rift valley, the tropical forests were by no means the only habitat available for terrestrial animals such as the therapsids, the largely quadrupedal predecessors of the primates and most other mammals. Even the tropics harboured open ground at riverbanks, rocky outcrops and along coastlines as well as arid regions to the north and south. So we can be certain that, all along the divergence of the therapsid line there would be animals that occupied all these terrestrial niches. With the proliferation of lower level bushes and fruit trees made possible by the great disruption of the forest canopy we can be sure that these new opportunities were exploited to greatly extend diversification of ground-based therapsids.

The opening of grasslands, perhaps aided by a chance chromosome doubling event, also spurred the development at the forest fringes of flowering plants such as the rosaceae, bushes and small trees with abundant fruit at low levels. Providing yet another niche. One that would favour bipedal adaptations of some of the therapsid stock.

We have plenty of examples of previously obligate quadrupeds animals who occupy such niches today. The bear is one such example. Its bipedal option greatly facilitates harvesting of fruits, small animals and honey. Tripling its vertical feeding range. Other well-known selective advantages of standing up, particularly in open country, include improved surveillance for prey or predators. Also for offence and defence (even a cat will rise up on its rear quarters when cornered to increase the vertical range of its claw weaponry), or for wading in deep water. The latter capability eventually being of particular significance for our own predecessors.
Furthermore, within the primate order itself we find that habitual bipedality, or at least vertical orientation of the upper body is not at all uncommon. So, in many ways, although primates have adapted to a range of different lifestyles, our species is in many ways just "more of the same".

An interesting aspect of the mindset which is locked in by tradition is the idea that, because we have arboreal cousins among the great apes, our ancestors must have "come down from the trees". It is overlooked that existing cousins of the cat and the bear are truly arboreal. Among the cat family are clouded leopard of Asia, the margay, which inhabits the tropical rain forests of the Americas, and the marbled cat, which has found its niche high in the forest canopies of the Himalayan foothills and in Sumatra and Borneo. From the other great division of the Carnivora, we find the wolf, very ill-adapted for tree climbing and the seal is even more so.

But among the bears, which also fall into this sub-order, we find the truly arboreal Andean spectacled bear. But in the case of animals such as cats, dogs, bears and seals, and in spite of the fact that almost all members of the cat family are superbly adapted for tree climbing, we have no equivalent evolutionary myth of these creatures having "come down from the trees". A conjecture which would likely be met with derision. But nature is very wonderful and exceedingly tricky.

For we find here a glorious irony. For, while there is no strong evidence of an arboreal ancestry for humans, some very recent fossil studies by a Belgian team led by Floréal Solé suggest that the common ancestor of the Carnivora does seem to have been arboreal.

*Art by Charlène Letenneur and Pascale Golinvaux*

Quite apart from parsimony, which would indicate that the most straightforward route to bipedalism is the most probable, there is a line of evidence which gives that model some substance.

This concerns the eccrine sweat glands. The original function of these devices seems to be to have been that of providing a degree of adhesion for the feet. To help give a good grip. They are common among mammals. Among the primates, however we find an interesting exaption in which these glands are found, to varying extents, over the body. Where they perform a very different function. That of cooling by evaporation. A function for which we would not expect any natural selection pressure for tropical rain-forest dwellers. Sweat glands on an animal living in conditions of high relative humidity making about as much sense as tits on a bull. So we can be pretty sure that this adaptation arose in the more arid environments, the province of ground-dwellers.

Sure enough, if we look at the distribution of primates that employ this cooling feature we find the prosimions, and within our own simian ancestral category, the rhesus monkey, the patas monkey, the stump-tailed macaque, the Japanese monkey, and baboon. All of which are primarily ground-dwellers that occupy open ground at riverbanks, rocky outcrops and along coastlines as well as arid regions to the north and south. Exactly the expected range of habitats mentioned above.

This early exaption, selected for through the ages by our ground based, not arboreal, ancestors has been mostly discarded by the great apes such as the gibbon and orang-utan that took to the rain-forest trees early. The ancestors of the three ape lineages which, five million years ago were to go their own evolutionary ways, retaining these somatic eccrine glands until this very day. For two of them, the chimpanzee and the gorilla, the five million years of disuse has not been enough to erase these sweat glands, but their redundant thermo-regulation function no longer works. Much as the useful capability for synthesising vitamin C was lost by our very remote, presumably frugiverous, ancestors, but most of the molecular machinery remains.

For the last member of the trio, of which the human is the only extant end-result, the utilization of the eccrine glands was, along with its obligate bipedalism, what made possible the adaptation to the semi-nomadic endurance-hunter with the ability to operate within a wide range of environments.

Yet another seriously flawed argument for our arboreal ancestry is our possession, in common with most other primates, of an opposable digit. This, it is affirmed, derives from the ability to grasp tree branches for climbing. In fact, the reverse is far more likely. That the use of prehensile fore paws with opposable digits to pick up small food items such as fruit and insects ***preceded*** the clawless tree-ward migration of various primate species.

That primates have nails rather than claws is, in itself strongly indicative of manipulation being the primary adaptation here. Today's squirrel has this feature, despite using its claws for climbing. Pandas, terrestrial animals which also sometimes stand up while foraging, have actually acquired a sixth digit for manipulation of food. In fact, it appears that the "invention" of opposable digits for this purpose harks right back to the theropod dinosaurs.

It might be argued that at least the prehensile feet of the gorilla and chimpanzees are suggestive of an arboreal rather than a ground based ancestry. After all, the feet and supportive nervous systems are configured this way. But because the gorilla, although mostly ground based, spends on average 10% of its time above ground and the chimpanzee around 50% the respective levels of pedal prehensility in these species might well be attributable to convergence rather than relics of ancestry. Since convergence usually results in closely analogous rather than exactly similar features, comparisons of the relative morphologies and functions of the feet of gorillas and chimps against those of the "older", tree-based great apes could possibly throw some light on this issue.

# APPENDIX B

## THE LANGUAGE OF SIMPLICITY

Cantonese is to myself, and to anybody who hasn't yet learned it, a horrendously difficult language with bizarre symbolism.. Yet hardly any would attribute a super-high "intelligence" to its users. The Chinese might indeed be intrinsically smarter than the English speaking world, but I seriously doubt that there is any measurable difference at all. Also, I am quite confident that, given quite a lot time, I could learn to understand Cantonese. So I find it rather remarkable that so many folk inappropriately stand in awe of mathematics.

A language that, compared with Cantonese, is very lean in symbolism, has a far, far smaller "alphabet" and certainly employs a much more straightforward and logical "word" structure and grammar than the English language.

One reason is perhaps that mathematics has long-standing mystical and platonic associations. This is evident from the earliest times when arithmetic and geometry were intimately entwined with numerology and astrology. Resurfacing from time to time with such incarnations as the ideology contrived by the Pythagoreans, Kepler's revival of the "music of the spheres" and today's string theory fad.

Another reason is that, because its direct correspondence with real world situations is limited to adult situations such as commerce and mensuration, the infant is not motivated (or is seldom even given the opportunity) to engage with mathematics in those first five or so years when its language-acquisition potential is at its peak. So it tends to be perceived as a "strange" language.

Yet again, as with music and the graphic arts, the diversity of intrinsic talents endowed to individuals by nature allows some brains to be configured in a certain way that makes numerical calculation very easy. Some of the most striking examples are found among what used to be called "idiot savants" but are now more kindly, and more accurately, referred to as autistic-spectrum prodigies. Some of these individuals have quite astounding memorization and number-crunching capabilities. But, sadly, this extraordinary degree of neural specialisation often comes at the expense of qualities which the majority of us take for granted. Just as it is wrong to think in terms of "brilliant minds" it is wrong to classify some as idiots. Nature experiments with all manner of configurations and we are each the result of some such experiment. However, those for whom the skill of conscious calculation comes easy have, all other things being equal, a significant motivational advantage for engaging with the language of mathematics. And from there exploring the far wider and more intriguing aspects of the subject.

The alert reader may have wondered why I have used the term "conscious calculation" in the previous paragraph. There is an ambiguity here that stems from the rather tenuous distinction that can be made between "being aware" and "being aware of being aware". The latter condition being almost certainly unique to our species and a direct consequence of the sharing of imagination called language. There is absolutely no doubt that the bulk of computation performed by our nervous system is carried out by processes that we usually designate as "the subconscious". Also that there are aspects of the subconscious that are capable of enormously higher computation speeds than those used in the rather ponderous processes of arithmetic that are conducted within the domain of the primary focus of awareness. In "Unusual Perspectives I wrote:

> *"For the brum rather ponderously handles its navigation strategies in the external world while always interacting cooperatively with bellum , the lightning-fast bodily tactician. The footballer's brum brum plans a route down the field and calls for mid-course corrections, the bellum handles the running and the twists and turns, the brum selects the goal area and the bellum kicks the ball. The brum notes the outcome and, using the dictionary and grammar rules list stored in its memory encodes an appropriate pattern of impulses which, when received and actioned by the bellum, produce a standard sequence of muscular actions that correspond to a vocalisation of triumph or invective accordingly.*
>
> *The contrast between the computational speed of the cerebellum and that of the brum is great indeed. Imagine the time it would take, assuming you had enough grounding in applied mathematics to start with, to use your brum to perform all the calculations needed to determine the end of the trajectory of a cricket ball, or base-ball from strike to catch. Now the bellum might cheat a bit with look-up tables embedded in its memory banks, it is true, but nevertheless the race will be well over (and the catch missed) while your poor old brum is chugging away at the first steps. Because that is not the sort of task that brums are cut out for.*
>
> *The brum of no other animal does this sort of thing. It is only the presence of elaborate Imaginatory "software" embedded within the memory banks of the brum which enables us to muddle along slowly with a bit of arithmetic at all. Mind you, there are a few people who can do a lot better than others in this respect, although not in the same league as the bellum. This may be just that they wound up with better software for some reason or perhaps they have some kind of freak connectivity. But it is also interesting to*

*speculate (perhaps a bit wildly) that they might in some way perform their feats by some form of direct access to the bellum. "*

New light has been thrown upon this model by experimental work of the kind mentioned in chapter 25, which shows that the neurons in the cerebellum are ***five times*** more numerous than in the cortex. This not only lends weight to the generalities of the above model but also leads to the speculation that the cerebellum may play an even greater part in the overall process, perhaps even being directly involved in some aspects of imagination.

While an innate propensity for fast calculation may provide motivation for a learner, and often helpful to the professional, by itself it doth not a good mathematician make. Any more than a winner of a spelling bee has what it takes to be a creative writer. Or the ability to be able to play many different scales necessarily makes for a creative musician. The higher levels of all arts are not engendered by arithmetical ability or by memory skills but rather by the exercise of imagination. That the language of mathematics is indeed an art is expressed by G.H. Hardy in *A Mathematician's Apology*:

*"I have never done anything 'useful'. No discovery of mine has made, or is likely to make, directly or indirectly, for good or ill, the least difference to the amenity of the world."*
*"The mathematician's patterns, like the painter's or the poet's must be beautiful; the ideas like the colours or the words, must fit together in a harmonious way. Beauty is the first test: there is no permanent place in the world for ugly mathematics."*

*"I am only interested in mathematics as a creative art."*

That is, of course, just one aspect of mathematics. Just as is the case with the natural languages, it can also be used to generate fictions. In fact it is largely based on fictions. The point, the two dimensional straight line have never been observed in the common sense world, or the world of the of the scientist, who relies on empirically derived information. Even the simplest concepts of number involve inexact representations. Two pebbles on the beach may look very similar but are never exactly the same. In fact it is only at atomic levels that we can say that maybe, in principle, there are separate entities that could be identical. But, even then, as Heisenberg pointed out, we can't be sure, for it is impossible to pin down all their properties at once. Infinities, too, are inherently unobservable and, although the concept can be useful, it can also lead to paradoxes within maths itself.

Another class of fictions which happens to be popular among some of the more vocal, if perhaps less astute theoretical physicists of the day, is that of the existence of spacial dimensions greater than three. For which there is not a whit of evidence, either direct or even incontrovertibly circumstantial. And, of course, within mathematics itself, you can have any old damn number of dimensions you choose! Try as you might you curl them up so they are invisible to our eyes and undetectable by science. Very reminiscent of the ghosts, spirit world and angels dreamed up within the natural languages.

The natural languages can be deployed as art for art's sake, or to generate fictions, some of which may occasionally lead to new insights into the real world, or fulfil their most common role, that of utility. Apart from works of art, the billions of artefacts that, with our involvement, language has effectively generated over the years, are characterised by their utility. And this is equally true, as we might expect, of the formalised subset of language that is mathematics. Languages of both kinds are basically utility-driven. While aesthetic spin-off is bounteous, inevitable and enjoyable it is utility that has provide the evolutionary impetus, the novelty, the selection pressure and selection.

There can be no question but the language of mathematics ,with its early roots found in such activities as mensuration, commerce and gambling, has become a major player in most aspects of everyday life and an indispensable tool in science. Particularly in physics, where the relatively straightforward phenomena under study lend themselves to manipulation within that simpler and more precise language. In the words of Bertrand Russell:

> *"Physics is mathematical not because we know so much about the physical world but because we know so little."*

Producing models that not only enhance understanding but also allow calculation from generalities and predictions of experimental or observational outcomes which can be tested by the scientist.
Einstein famously observed:

> *" As far as the laws of mathematics refer to reality, they are not certain; and as far as they are certain, they do not refer to reality."*

It is hard to deny the truth of this statement. But the real point is that, as long as we keep things in perspective, it doesn't really matter. For it is not only within the domain of mathematics that we find this perceived weakness. Although, camouflaged by their inherently very diffuse nature, far less obvious, it equally applies to the natural languages. The underlying issue is really that all mathematics is metaphor. But so, too, are our imaginations. They are, so the empiricist believes, neural representations of an external reality. Even if that external reality, is, as we assume, valid, the neural configurations are proxies. They are still proxies if represented in writing, or on flash drive, or in any other form. Thus ***all*** language, including that of mathematics, must properly be regarded as metaphor.

While maths is invaluable for helping to understand the simpler phenomena, it is generally at a loss when confronted with the great abstractions and much higher levels of complexity found in sciences such as biology. Try giving a mathematical description of a tree or a rhinoceros or even a bacterium! It is only in the foothills of subsidiary processes that the simplicity of the language can play its part in clarification. It is important to be able to distinguish those cases where it is out of its league.

The aspects of mathematics that most consistently corresponds to complex “real life” situations, even in the intricacies of biology is statistics. Indeed, it was within the complexities of the human network that it was born. In gambling. And it has turned out that gambles, probabilities, are a large part of what makes our universe tick. At ***all*** levels! Biology included.

In saying this, I am not advocating attempts to find exact definitions of words. For by its very nature, natural language lacks precision. This is because of the very high level of complexity and abstraction in even seemingly simple concepts. Even within the simplified and very formal coding of derived mathematics there lie peripheral ambiguities and complete precision is not always achievable. Nor is it generally even desirable. For words reflect aspects of imagination and imagination has an inherent (and very necessary) high degree of overlap and redundancy.

The word "tree" represents an extremely high level of abstraction, covering a vast range of species and individual configurations a well as the many non-biological descriptions of systems characterised by branching structures. (We will ignore such trivial and seemingly unrelated uses such as the "shoe-tree" here) Nevertheless, it is an important word of great utility which certainly conveys meaning.

John von Neumann once told a worried student "Young man, in mathematics you don't understand things. You just get used to them." This is basically true not only of mathematics but also of the more comprehensive natural languages. We “just get used to” concepts and their manipulation, starting from when we are babies. It is the process we call learning. As with natural languages, achieving fluency depends largely on motivation, although natural talent naturally also plays a part. The most highly motivated budding linguists, then, are new-born children. But for those early years the simplification of mathematics (other than the concept of number) are not of much use.

# APPENDIX C

## WYSIWYG (aka SCIENCE)

After exploring the philosophical landscape in my younger days, I eventually came to the conclusion that empiricism, on which technology and science are founded, is the only "faith" to be both meaningful and useful. For the greater part of philosophy, pretty well "anything goes", even solipsism, the lonely belief that there is no external reality,

Previously called "Natural Philosophy", it is unique among the philosophical schools of thought in that it's spirit is captured is captured in the acronym WYSIWYG. What you see is what you get! A way of thought that started to separate out from mystical world-views in the time of Locke and Hume.

Science is based in empiricism which has, as its fundamental tenet, the interpretation of nature based on observational/experimental evidence. In stark contrast, all spirituality and mysticism is based on tradition, hearsay and mere speculation. So there is a fundamental incompatibility between these ways of thought. Any belief system, including the "scientific" inherently requires faith in its axioms. The main axiom of empiricists (I am of that faith) is that an external reality exists and that the world, for the most part, is properly represented by the chemical, mechanical, thermal and visual inputs to our various sensory receptors. An acceptance that the information presented to us by our senses are genuine representations of the external world.

There are, of course, exceptions. These false readings, illusions, arise from the extensive processing which takes place between the input of raw data in detection of signals and their mental interpretation. Distortions (and omissions) which, in either our evolutionary or developmental past, have proved to optimize behaviour in ordinary circumstances. Adaptations that have evolved to best fit our interaction with the environmental network. Nature's equivalent to "constructive accountancy"

One of the most remarkable commonplace deceptions is that provided by the moon. Most spectacularly when full, it appears to be very large when rising above the horizon but shrinks to a much small size as it approaches its zenith. This is very convincing. It is not some kind of refractive effect resulting from viewing through a thicker layer of atmosphere. It is quite easy prove to yourself that the apparent difference in size is not real, but all in your mind. Simply take a cardboard tube, hold it to your eye and look through it at the rising moon. Trim it to size so that the moon appears to just fill the hole at the end. Check again when the" tiny" moon is high in the sky: Voilà!

This particular kind of filtering has arisen as a result of our need to apply perspective compensation when viewing terrestrial objects at a distance. There are reports of tribes who have been confined to dense jungle habitats who are gob-smacked by the " tiny animals" they see when first introduced to the Savannah. That the moon illusion is a learned rather than a genetic phenomenon is supported by research that suggests that townies are more susceptible to it than those who live in the wide open spaces.

But the reasons for, and lack of validity of, illusions such as these is revealed by more sophisticated analysis of the primary sensory inputs. And we have far better devices than cardboard tubes with which to make such checks. Today, greatly extended in range, selectivity and sensitivity by the multiplicity of external transducers at our service. For example, the microscope, the telescope, the spectroscope. Not only for gathering new data but also to correct for the biases inherent in our neural processing.

Furthermore, we can usually explain the function of superficially "wrong" inputs in terms of learned behaviours or evolutionary adaptive pressures.

Empiricism, in its embodiment as science, differs from all the other philosophies in that it, in synergy with the technologies with which it is inextricably intertwined, uniquely produces artefacts. Bridges, motor vehicles, jumbo-jets to mention but a few. Of course, a really die-hard solipsist will deny the actuality of these entities. But the great irony is that those who embrace philosophies other than empiricism are quite happy to put their faith in the use of these artefacts. A wonderful inconsistency! Often using a complex product of empirically derived technology to purvey a negation of the very materialism which has generated it. So, judging by all the artefacts and technologies it has produced, empiricism works pretty well. At any practical level, the sciences and their technological symbionts are almost universally trusted. The worlds religious leaders show much greater faith in aerodynamics than in their much vaunted "power of prayer" for transcontinental excursions.

Most of those who openly subscribe to superstitious interpretations of the world around them are at least easy to recognize. More insidious are those purveyors of fantasies that, disguised as *bona fide* science, hit the headlines with monotonous regularity. Managing to fool swathes of quite rational individuals with extensive bodies of pseudo-scientific supposition. Speculations that not only have no factual basis or consonance with the main body of science but are, for the most part essentially untestable. And it is not at all uncommon for many such to employ the facilities kindly provided by science and technology in their pursuit of metaphysical moonbeams

The field of theoretical physics is remarkably susceptible to misrepresentations of this kind. It is particularly evident among the glitterati who aggressively promote as fact unevidenced constructs such as string theory, M-theory and the like. Such individuals are often acclaimed by the general public as "great scientists". Which is very far from the truth. Indeed, being mere mathematicians involved in speculation, they cannot justifiably lay claim to being scientists at all. Far better described as science fiction writers who primarily use the language of mathematics for the composition of their speculations. While mathematics is an enormously valuable tool in the service of science there is no guarantee that any of the constructs of that language correspond to the real world. Exactly as with the natural languages, internally consistent fictions can flourish.

This group is heavily influenced by the kind of thinking put forward by the ancient Greek philosopher, Plato. Among whose strange pronouncements we find that "reality is unavailable to those who use their senses" and that "he who sees with his eyes is blind" . Today this view is expressed as the belief in such things as "abstract objects", objects that do not exist in space or time and are "non-physical and non-mental". So what can we say to get a sensible handle on the nature of science? Carl Sagan, who was remarkably sagacious and articulate in these matters, observed:

> *"There is no other species on the Earth that does science. It is, so far, entirely a human invention, evolved by natural selection in the brain for one simple reason: it works.*
>
> *It is not perfect. It can be misused. It is only a tool. But it is by far the best tool we have, self-correcting, ongoing, applicable to everything. It has two rules.*
> *First: there are no sacred truths; all assumptions must be critically examined; arguments from authority are worthless.*
> *Second: whatever is inconsistent with the facts must be discarded or revised. We must understand the Cosmos as it is and not confuse how it is with how we wish it to be."*

# The evolution of science

*"Snowblindedness", faulty integration, and other forms of the individual pathology of thoughts, are reflected on a vastly magnified scale in the history of science: and the power of habit over the individual mind is reflected in the conservatism of scientific bodies and schools – which has impeded progress for periods ranging from years to centuries.*
***Arthur Koestler***

Thomas Kuhn, in "The Structure of Scientific Revolutions" draws attention to the punctuated advancement of science, particularly within the in the traditional academic setting.. He points to the fact that most research is conducted within a framework represented by established textbooks and methods. The results of this work providing rather slow incremental extensions of this framework. This works well for a while as the model is slowly fleshed out.

But eventually, as a result of new instruments becoming available, new methods, certain "rogue" researchers (sometimes outsiders) or sheer serendipity, cracks start to appear in the edifice to the extent that a major revision, an essentially new framework is required. This drastic revision characterized by a rapid rate of change until the new paradigm, complete with textbooks and methods, becomes flavour of the day.

To some extent this historical account lacks relevance to modern times and also smacks of its "Ivory Tower" origin. The very productive "blue sky" research policies implemented by Bell laboratories in the middle of the twentieth century and the provision of comparable commercial work environments by enterprises such as Google have changed the situation. With real time access to current understandings in all fields, and the formation interdisciplinary networks, the death of this cyclic phenomenon could be imminent.

Be that as it may, despite the underlying incremental advance in intricacy, drastic changes of rate are characteristic of evolutionary processes in general. Jay Gould's advocacy of "punctuated evolution" in the biological field similarly pointed out that, although basically occurring in tiny steps, temporal discontinuities often occur. These can be attributed to having to wait for the environmental conditions that enable "progress" to eventuate. Translated to the development of science this corresponds to crises requiring major revisions of the established intellectual environment. Whereas, in techno-evolution, market economics and conflict are found to be among the more crucial considerations.

That is not to say that passivity is sufficient for progress. Much of what Kuhn calls "normal science", an activity more akin to that of a hunter rather than a gatherer, can be laborious, for not all knowledge comes easy. The data miner needs to pro-actively prise out and glean many of nature's secrets.

# The nature of science.

**The tools of science.**

Scientific method: This is a concept widely flaunted in elementary text-books and by popular science journals and is frequently even used in vague definitions of science. This is very wrong There is no slick prescription for its practice.
Rather it is characterized by the use of various tools.

- Observations, direct or indirect. These are of the essence But the recording of lists of observations is, in itself, not very productive.
- Taxonomy. The sorting of observations into various kinds of category.
- The search for pattern. The search may be newly acquired observational data, or that obtained from the resources

provided by taxonomy.

- Identification of pattern. Hypothesis formation.
- Validation of perceived pattern by controlled experiment. This is an ideal in which all variables other than the one under study are fixed. However, this is not always possible. Indeed, sometimes experiment is impracticable.
- Validation of perceived pattern by consilience. The pattern is checked for compatibility with the established main body of science.
- Exploration. In the traditional sense, say, exploring the ocean deeps. Or the metaphorical sense of open-ended experiment. Jus' foolin' around.

These tools being used in any combination and not always in any particular order. They provide no certainties, Simply our best current guess.

The much vaunted controlled experiment has a significant role in the practice of science but it is primarily that of validation and only results in discovery in the event that something goes wrong. As reflected by Isaac Asimov's comment:

> *" The most exciting phrase to hear in science, the one that heralds new discoveries, is not 'Eureka!, but 'That's funny...'"*

Much of the research carried out today, particularly in the medical field, is mere screening. No hypothesis is involved. An example is the checking of huge numbers of marine organism for anti-cancer compounds. Statistically increasing the chance of discovery.

Open-ended experimentation, "pure" research, historically dominated by those few endowed with both wealth and curiosity, is in these times proportionately less common, since having no pre-specified outcome, it is usually hard to fund. For, As Albert Einstein pointed out in this regard:

> *"If we knew what we were doing then it wouldn't be research, would it.?"*

And, still, many serendipitous discoveries originate in technology and are fed back into the loop.

There is a popular misconception that scientific methodology is simply an iteration of a hypothesis - experiment - conclusion routine. Fortunately that is not the case. Science would be a rather dull field of endeavour were it so. And far less productive.

Another popular fad spread through the class-rooms is the naive belief that falsification is the be-all and end-all of validation. Karl Popper's contribution to the philosophy of science was indubitably worthwhile. But, just as Karl Marx famously denied bring a "Marxist" and Charles Darwin denied bring a "Darwinian", I suspect that if Popper were alive today he would not admit to being a "Popperian" For falsifiability alone doth not a sound hypothesis make. For one thing, a hypothesis can often be re-framed as its inverse. But, more importantly, an unlimited supply of falsifiable hypotheses can be generated ad hoc.

For example, consider the following:

> *"If a mixture of equal proportions of cerium, gallium, carbon and cat's urine are blended and placed in a porcelain vessel they will, under the specified conditions, be transmuted into a quantity of gold having 2.73 times the combined initial mass. The specified conditions are that the sun shall be within 18 degrees of its zenith on that day in 2040 when the major planets are next in conjunction and the mixture is exposed to a magnetic field between 1.07 and 247.3 gauss"*

An eminently falsifiable hypothesis! But is it sound? Is it useful in any way? Is it worth pursuing? Does it have any place in the domain of science? You may draw your own conclusions.

**(Warning: Don't try falsifying this at home, kids!)**

Neither testability nor falsifiability alone are sufficient criteria for winnowing the silly from the sound. They are not entirely unimportant but consilience is a far better guide.

## Scientists

Opinions of what kind of individuals are included in this category varies enormously. In the minds of most folk, "the man in the street", "Joe Six-pack" (and the female equivalents), for instance, it seems that anybody who has a job in a laboratory will fit into this perceptive pigeon-hole. Or, perhaps, anybody having a science degree. Or even that subset of mathematicians that publicly promote pseudoscientific fantasies. At the other extreme of the spectrum will be found the ivory towered variety of academic who restricts the description only to postgraduates engaged in pure research. Those of the cognoscenti who happen to have a better hold on reality just accept that the distinctions are blurred and generalizations of no consequence.

In terms of human qualities, those involved with the sciences in any capacity at any level, from routine quality control or sampling to the professor emeritus or Nobel laureate, simply reflect the flaws found in the general population. They can be, and quite often are, mean, deceitful, daft, dogmatic or arrogant. And some, despite their exposure to empirical ways of thought, still cling to superstition or extreme contrarian dogmas. Thomas Malthus observed:

> *"It is an acknowledged truth in philosophy that a just theory will always be confirmed by experiment. Yet so much friction, and so many minute circumstances occur in practice, which it is next to impossible for the most enlarged and penetrating mind to foresee, that on few subjects can any theory be pronounced just, till all the arguments against it have been maturely weighed and clearly and consistently refuted"*

For some, perhaps the majority, the pursuit of science is simply a way of earning a living. And it is from such journeymen that most of the results of what is often tedious, repetitive and painstaking work emerge. These folk carring out all the groundwork in their own small fields perhaps qualify as the true unsung heroes of science and technology. For, in this era of specialisation, very little advance would be made without them. Others get bogged down with administration and teaching. In the middle are those driven by great enthusiasm who beaver away and provide new insights by avid recombination of ideas. Many of whom also often go largely unrecognised and frequently have their achievements unfairly usurped by unscrupulous seekers of tenure, power, recognition or commercial gain. The Nobel trail, for example,, shows many instances of these. Then there are the glitterati of the establishment, many of whom have little of real sustance to say but make an awful lot of noise saying it. For fame is certainly not proportional to ability or creative thought. As noted by science historian Scott Weingart in his interesting blog " The Scottblog Irregular":

> *"...it's a fairly widely-known phenomenon that in science, famous names are more likely to be published than obscure ones (there are many anecdotes about anonymous papers being rejected until the original, famous author is revealed, at which point the paper is magically accepted). The number of accepted papers may be as much a proxy for fame as it is for performance, so the results do not support their hypothesis. The authors then look at awards given to actors and writers, however those awards suffer the same issues: the more well-known an actor, the the more likely they'll be used in good movies, the more likely they'll be visible to award-givers, etc. Again, awards are not a proxy for the quality of a performance. The paper then goes on to measure elected officials based on votes in elections. I don't think I need to go on about how votes might not map one-to-one on the performance and prowess of an elected official."*

The proudly boasted process of peer review is similarly open to question. Commenters such as Daniel Kahneman point out that a worryingly high proportion of published papers today contain results that are not replicable. One factor is the making of extravagant claims can greatly enhance the chances of getting a research grant. As can the practice of including terms not essentially related to the work in hand. An article in The Economist Oct 19th 2013 remarks:

> *"Little wonder that one in three researchers knows of a colleague who has pepped up a paper by, say, excluding inconvenient data from results "based on a gut feeling". And as more research teams around the world work on a problem, the odds shorten that at least one will fall prey to an honest confusion between the sweet signal of a genuine discovery and a freak of the statistical noise. Such spurious correlations are often recorded in journals eager for startling papers. If they touch on drinking wine, going senile or letting children play video games, they may well command the front pages of newspapers, too."*

Further clouding the sky is Massimo Pigluucci's comment in his excellent book "Nonsense on Stilts" to the effect that:

> ***"Every scientific theory proposed in the past has eventually been proven wrong and has given way to new theories".***

Not too far from the truth, although "inadequate" rather than "wrong" would preferable in many cases. Newtonian mechanics being in the former class and the phlogiston theory in the latter.

So are we to believe that science is of little worth and in disarray?

No, for the faults lie in the inadequacies of the poor snoutless apes who are its sole practitioners. Saddled with the multiple variations upon their genetic and cultural histories and their place within the biospheric network.

In November 1947 Winston Churchill, always remarkably insightful for a political figure, remarked:

> *"Many forms of Government have been tried, and will be tried in this world of sin and woe. No one pretends that democracy is perfect or all-wise. Indeed it has been said that democracy is the worst form of Government except for all those other forms that have been tried from time to time"*

In a similar vein we can  similarly assert that although the ways and practice of science may be far from perfect they are the best we have.